Praise for *Forensic*

"A gripping history . . . [that] contains a wealth of surprising information . . . If McDermid is ever stuck for inspiration for her novels she could do worse than turn to her own book of the dead for inspiration."
—*Independent* (UK)

"The crime novelist burrows into the wild history of forensics, interviewing top scientists and culling from innovative research to explain practices like using maggots to calculate time of death."
—*Entertainment Weekly* ("Brainy & Brilliant Beach Books")

"McDermid shows . . . laser-like curiosity, grisly humor, and consistent compassion . . . Armchair Clarice Starlings raised on *Sherlock*, *CSI*, and *Dexter*, will devour every maggot mosh pit and bloody fingerprint."
—*Bust*

"Gruesomely fascinating . . . Fans of McDermid's fiction will gain a greater understanding of where her ideas come from."
—*Publishers Weekly*

"McDermid has not lost her early journalistic genius for telling a good story plainly and with passion."
—*Times Saturday Review* (UK)

"A guide to criminal forensics that is every bit as compelling as the best of the fictional genre."
—*Irish Times*

"[McDermid] approaches the grisly realities of crime scenes and corpses with a neophyte's sense of wonder . . . A satisfying insider's excursion."
—*Kirkus Reviews*

"McDermid would make a good doctor, managing to be clinically precise but engaging at the same time . . . A highly readable, eye-opening account of the way in which criminals have slowly had their wings clipped and their getaways thwarted over the past hundred and more years."
—*Herald* (Scotland)

"[*Forensics*] melds bite-sized history lessons and the reconstructed narratives of scientifically significant cases—both modern and ancient—with explanations of the technologies that were crucial in their prosecution . . . McDermid has an eye for authenticity."
—*Winnipeg Free Press*

"Every chapter is packed with facts that are by turns astonishing . . . and disturbing . . . Fascinating."
—*Tampa Bay Times*

Also by Val McDermid

A Place of Execution
Killing the Shadows
The Distant Echo
The Grave Tattoo
A Darker Domain
Trick of the Dark
The Vanishing Point
Northanger Abbey
The Skeleton Road

TONY HILL NOVELS

The Mermaids Singing
The Wire in the Blood
The Last Temptation
The Torment of Others
Beneath the Bleeding
Fever of the Bone
The Retribution
Cross and Burn
Splinter the Silence

KATE BRANNIGAN NOVELS

Dead Beat
Kick Back
Crack Down
Clean Break
Blue Genes
Star Struck

LINDSAY GORDON NOVELS

Report for Murder
Common Murder
Final Edition
Union Jack
Booked for Murder
Hostage to Murder

SHORT STORY COLLECTIONS

The Writing on the Wall
Stranded
Christmas is Murder

NON FICTION

A Suitable Job for a Woman

FORENSICS

WHAT BUGS, BURNS, PRINTS, DNA, AND MORE TELL US ABOUT CRIME

VAL McDERMID

Grove Press
New York

First published by Profile Books in partnership with Wellcome Collection in the UK in 2014. Wellcome content is copyright to Wellcome Collection, 2014. Grove Atlantic thanks the Wellcome Collection for its support.

Published simultaneously in Canada
Printed in the United States of America

ISBN 978-0-8021-2515-6
eISBN 978-0-8021-9105-2

Grove Press
an imprint of Grove Atlantic
154 West 14th Street
New York, NY 10011

Distributed by Publishers Group West

groveatlantic.com

21 22 23 10 9 8

CONTENTS

For Cameron, with love

Without science, there would be no you;
without you, the future would offer a much narrower prospect.
Good stuff, science.

PREFACE

The face of justice we know today has not always been judicious. The notion that the criminal law should be based on evidence is a relatively recent one. For centuries, people were accused and found guilty because of their lack of status; because they weren't from round here; because they or their wife or their mother was handy with herbs; because of the colour of their skin; because they had sex with an inappropriate partner; because they were in the wrong place at the wrong time; or just because.

What changed that was the growing understanding that the scene of a crime held all sorts of useful information and that branches of science were emerging which could read that information and present it in a courtroom.

The trickle of eighteenth-century scientific discoveries that turned into a flood in the nineteenth century and beyond soon found practical applications far beyond the laboratory bench. The idea of proper criminal investigation was just beginning to take hold, and some of those early detectives were eager to find evidence that would back up their theories of the crimes they were investigating.

Forensic – meaning, a form of legal evidence – science was born. And it soon became clear that many branches of scientific inquiry would have something to contribute to this new methodology.

One of the earliest examples brought together pathology and what we would now call document examination. In 1794, Edward

Culshaw was murdered with a pistol shot to the head. Back then, pistols were muzzle-loaded and a wad of crushed paper was tamped down to secure the balls and the powder in the gun. When the surgeon examined the body, he found the wad inside the head wound. He opened it up and it was revealed to be the torn corner of a ballad sheet.

The murder suspect, John Toms, was searched. In his pocket was a ballad sheet, the torn corner of which matched exactly the pistol wad. At his trial in Lancaster, Toms was convicted of murder.

I can only imagine how exciting it must have been to see these developments making the law a more certain instrument of justice. The scientists were helping the courts turn suspicion into certainty.

Take poison, for example. For hundreds of years, poison had been the murder method of choice. But, without reliable toxicology tests, it was almost impossible to prove. That was about to change.

But even in the earliest stages, that scientific evidence was open to question. In the late eighteenth century, a test had been devised to detect arsenic, but only in large quantities. Later, the test was refined to become more effective, thanks to the British chemist James Marsh.

In 1832, the prosecution called on him as an expert witness in chemistry during the murder trial of a man accused of poisoning his grandfather using arsenic in a cup of coffee. Marsh had conducted his test on a sample of the suspicious coffee and had demonstrated the presence of arsenic. But when he came to show it to the jury, the test sample had deteriorated, the result no longer clear. The accused walked free on the grounds of reasonable doubt.

But this wasn't a setback for the fledgling experts. James Marsh was a proper scientist. He regarded this failure as a spur towards success. His response to the embarrassment of his court appearance was to devise a better test. His definitive test was so effective it could detect even a tiny trace of arsenic; it was ultimately responsible for the hanging of many a Victorian poisoner who had not reckoned on forensic science. And it's still in use today.

The story of forensic science, of that road from crime scene to courtroom, is the stuff of thousands of crime novels. The application of science to the solving of crime is the reason I am gainfully employed. Not for the obvious reason – that forensic scientists are generous with their time and their knowledge – but because their work has transformed what happens in courtrooms all round the world.

We crime writers sometimes like to claim our genre has its roots in the deepest recesses of literary history. We claim antecedents in the Bible: fraud in the Garden of Eden; fratricide by Cain of Abel; the manslaughter of Uriah by King David. We try to convince ourselves that Shakespeare was one of us.

But the truth is that crime fiction proper only began with an evidence-based legal system. And that is what those pioneering scientists and detectives bequeathed us.

It was clear, even in the early days, that while science could help the courts, so the courts could push scientists on to greater heights. Both sides have a key role to play in the delivery of justice. For this book, I have spoken to leading forensic scientists about the history, the practice and the future of their disciplines. I have climbed to the pinnacle of the highest tower in the Natural History Museum in pursuit of maggots; I have been transported back to my own confrontations with sudden violent death; I have held somebody's heart in my hands. It's been a journey that has filled me with awe and respect. The stories these scientists have to tell us about that often tortuous journey from crime scene to courtroom are among the most fascinating you will ever read.

And a firm reminder that truth is stranger than fiction.

Val McDermid

ONE

THE CRIME SCENE

The scene is the silent witness

Peter Arnold, crime scene specialist

'Code Zero. Officer in need of assistance.' It's the call sign every British police officer dreads. One grey November afternoon in Bradford in 2005, PC Teresa Millburn's broken words on the radio sent a chill round the West Yorkshire Police control room. Her message heralded a case that touched everyone in the police community. That afternoon, the fear that cops live with every day became a bleak reality for two women.

Teresa and her partner, PC Sharon Beshenivsky, just nine months into the job, were near the end of their shift in their patrol car, their task a roving, watching brief. There to intervene in minor incidents. To be a visible presence on the streets. Sharon was looking forward to getting home for her youngest child's fourth birthday party and, with less than half an hour till she and Teresa clocked off, it looked like she would be in time for the cake and the party games.

Then, just after half-past three, a message came through. A silent attack alarm linked directly to the police central control had been set off in Universal Express, a local travel agent's shop. The two women would be passing anyway on their way back to the station, so they decided to take the call. They parked up opposite the shop and crossed the busy road to the long single-storey brick building, its picture windows obscured by vertical blinds.

As they reached the shop, they came face to face with a trio of armed robbers. Sharon was shot in the chest at point blank range. Later, at the trial of Sharon's killers, Teresa said, 'We were a stride apart. Sharon walked in front of me. Then she stopped. She stopped dead – she stopped that quickly that I overstepped. I heard a bang and Sharon fell to the ground.'

Moments later, Teresa was also shot in the chest. 'I was lying on the floor. I was coughing up blood. I could feel blood running down my nose and blood over my face, and I was gasping for breath.' Yet she managed to press the panic button and alert the control room with those fateful words 'Code Zero'.

Peter Arnold, a Crime Scene Investigator (CSI) for the Yorkshire and Humberside Scientific Support Services, heard the code call on the radio. 'I'll never forget it. I could see the scene from the police station; it was literally just up the road. And suddenly there was a sea of police officers running up the road. I've never seen so many police officers running at one time, it was like a fire evacuation.

'At first I didn't know what was going on. Then I heard over the radio that someone had been shot, possibly a police officer. So I just ran too. I was the first CSI to the scene. I wanted to support the officers in terms of getting the cordons up, making sure we'd got the scene preserved, because it was very emotive at the time, as you can imagine, and we just needed to bring some order to that.

'I spent the best part of two weeks processing that scene. Some very long hours. I'd start at seven in the morning and wouldn't get home till midnight. I remember being absolutely exhausted afterwards, but at the time I didn't care. That will live with me forever. I'll never forget that scene. Not because it was so high profile but because it was so personal, because it was a colleague that had been murdered. The fact that Sharon was a police officer made her part of my family. Others who knew her were even more upset, but they all swallowed their grief and got on with the job.'

Police Officer Sharon Beshenivsky, who died after being shot at point blank range by a gang of armed robbers

'And we had some very good forensic results that really contributed to the case as well, not just at that scene but at the peripheral scenes as well: the getaway vehicles and the premises they went to afterwards.'

The men responsible for the armed robbery that left Sharon Beshenivsky's husband a widower, and her three children motherless, were later brought to trial and jailed for life. The conviction was mainly down to the work of CSIs and other forensic experts, people who find evidence, interpret it and eventually present it in court. We will follow that evidential journey in this book.

Every sudden violent death carries its own story. To read it, investigators begin with two primary resources – the crime scene and the body of the deceased. Ideally they discover the body at the scene; looking at the relationship between the two

Edmond Locard, who opened the world's first crime investigation laboratory, also coined the forensic scientists' watchword, 'every contact leaves a trace.'

helps investigators reconstruct the sequence of events. But that's not always the case. Sharon Beshenivsky was rushed to hospital in the forlorn hope that she might be resuscitated. Other fatally wounded people sometimes manage to make their way some distance from where they were attacked. Some killers move the body, either with the intention of hiding it or simply to confuse detectives.

Whatever the circumstances, scientists have developed methods that provide detectives with an array of information to read the story of a death. To make that story credible in a court of law, the prosecution has to show that the evidence is robust and uncontaminated. And so crime scene management has become the front line in the investigation of murder. As Peter Arnold says, 'The scene is the silent witness. The victim can't tell us what happened, the suspect probably isn't going to tell us what happened, so we need to give a hypothesis that explains what has taken place.'

The accuracy of such hypotheses has developed in tandem with our understanding of what it's possible to learn from the scene of the crime. In the nineteenth century, as evidence-based legal proceedings became the norm, the preservation of evidence remained rudimentary. The notion of contamination wasn't part of the reckoning. And considering the narrow limits to what scientific analysis could achieve, this wasn't such a big problem. But the limits expanded as scientists applied their increasing knowledge in practical ways.

One of the key figures in the understanding of crime scene evidence was the Frenchman Edmond Locard. After studying medicine and law in Lyons, he opened the world's first crime investigation laboratory in 1910. The Lyon police department gave him two attic rooms and two assistants, and from these cramped beginnings he grew the place into an international centre. From an early age Locard had been an avid reader of Arthur Conan Doyle, and was particularly influenced by *A Study in Scarlet* (1887), in which Sherlock Holmes makes his first appearance. In that novel Holmes says, 'I have made a special study of cigar ashes – in fact, I have written a monograph upon the subject. I flatter myself that I can distinguish at a glance the ash of any known brand, either of cigar or of tobacco.' In 1929, Locard published a paper on the identification of tobacco by studying ashes found at a crime scene, 'The Analysis of Dust Traces'.

He wrote a landmark 7-volume textbook on what he called 'criminalistics', but probably his most influential contribution to forensic science is his simple phrase, known as the Locard Exchange Principle: 'Every contact leaves a trace.' He wrote: 'It is impossible for a criminal to act, especially considering the intensity of a crime, without leaving traces of his presence.' It might be fingerprints, footprints, identifiable fibres from his clothing or his environment, hair, skin, a weapon or items accidentally dropped or left behind. And the converse is also true – the crime leaves traces with the criminal. Dirt, fibres from the victim or the scene itself, DNA, blood or other stains. Locard demonstrated the power of this principle in his own investigations. In one case, he unmasked a killer who appeared to have a solid alibi for his girlfriend's murder. Locard analysed traces of pink dust found among the dirt under the suspect's fingernails and proved that the powder was a unique make-up made for the victim. Confronted with the evidence, the killer confessed.

The influence of dedicated laboratory scientists continues apace. But without the initial fastidious work at the crime scene, science has nothing to work on. One unlikely pioneer of reading a crime scene like a narrative was Frances Glessner Lee, a wealthy Chicago heiress who founded the Harvard School of Legal Medicine in 1931, the first of its kind in the US. Lee constructed a series of intricate replicas of actual crime scenes, complete with working doors, windows, cupboards and lights. She christened these macabre dolls' houses 'Nutshell Studies of Unexplained Death' and used them in a series of conferences on understanding crime scenes. Investigators spent up to ninety minutes studying the dioramas and were then invited to write a report on their conclusions. Erle Stanley Gardner, the crime writer whose Perry Mason mysteries were the basis for a long-running TV series, wrote: 'A person studying these models can learn more about circumstantial evidence in an hour than he could learn in months of abstract study.' The eighteen models are still being used for training purposes more than fifty years later by the Office of the Chief Medical Examiner of Maryland.

While Frances Glessner Lee would recognise the principles of modern crime scene management, most of the details would be alien to her. Paper suits, nitrile gloves, protective masks – all the paraphernalia of modern CSI work has given it a rigour that early criminalists only dreamed of. Such was the rigour that was brought to bear on Sharon Beshenivsky's murder; a textbook example of investigators following every promising lead to its conclusion. As always, detectives relied heavily on the information supplied by the forensics team.

At the front line in this process are the CSIs. They start their career path on a residential development programme that gives them a grounding in basic skills and techniques for identifying, collecting and preserving evidence. When they return to their base, they are closely mentored while they build up on-the-job experience, starting with lower level crime and working their way up to more difficult cases as they acquire knowledge and skills. They have to provide a portfolio of evidence over time to demonstrate their competency.

We've seen enough crime scenes processed on TV. We all think we know how it's done; the white-coated professionals painstakingly photographing, bagging and preserving vital evidence. But what is the reality? What do CSIs really do? What happens after a body has been discovered?

Generally, the first police to arrive are in uniform. The decision on whether the death in question is suspicious rests with a plain-clothes officer of the rank of Detective or above. Once the detective has determined that it could be a homicide, the scene is preserved for the CSIs. The police withdraw, set a perimeter with cordon tape and start a scene log. Anyone who enters or leaves the scene is recorded, so every possible source of contaminated evidence is listed.

A lead detective takes charge of the investigation. All CSIs are accountable to them, and the buck stops with them. The lead detective decides what they need as far as CSI response and evidence collection, but the DA may later request another testing closer to trial.

Peter Arnold is a trim bundle of energy with the sharp eyes of a blackbird and an obvious enthusiasm for his job. An Area Forensic Manager in the UK, he oversees CSIs, advises detectives and coordinates scientific resources, and his unit serves four separate police forces. They are the UK's biggest scientific support service outside the London Metropolitan Police, with a staff of around 500. They work a 24-hour shift pattern to provide a round-the-clock service to detectives investigating every type of crime imaginable. The service is based just off the M1 near Wakefield, in a custom-built centre named after Sir Alec Jeffreys, the father of DNA profiling. It overlooks a manmade lake whose rural tranquillity is in sharp contrast to the cutting-edge science that happens inside the building.

'As soon as I get that first call, I start coordinating resources,' says Peter. 'If it's an indoor scene, there's not so much of a hurry because nothing's going to get snowed on, nothing's going to get rained on; it's now a sterile, preserved scene and we will deal with it in a more considered manner. But if it's an outside scene, it's the middle of winter and it's about to throw down with rain, I'll get staff to the scene immediately to recover evidence before it gets destroyed.'

Because the primary scene of Sharon Beshenivsky's murder was out of doors, on a busy street, preserving the evidence was a priority. But that wasn't the only concern for Peter and his colleagues. 'People think of a crime scene, singular. But often we can end up with five or six relevant scenes for a homicide – where a person has been killed, where the suspects have gone afterwards, a vehicle that the suspects have travelled in, where the suspects are arrested and, if the body is removed, where it ends up. All of those different scenes need to be processed separately.'

The first issue for CSIs working these scenes is security. It may be that someone's been shot and the suspect is still at large. CSIs don't wear stab vests, or carry guns, Tasers or handcuffs. They're not trained to arrest violent people, even though they are often dealing with scenes left by them. So, if necessary, armed officers are deployed to protect the CSIs.

After safety comes preservation. Peter explains, 'We might arrive at a scene where a house is cordoned off but the suspects have run up the road and got into a getaway car. If we've still got cars driving on that road they could be driving over bullets or bloodstains or tyre marks. So it would make sense to cordon off the whole street till we can collect the evidence.'

Once the cordon is in place the crime scene investigator puts on her full protective equipment: a white scene suit; a hairnet or hood; two pairs of protective gloves (because some liquids can seep through the first pair); and overshoes. She also puts on a surgical mask so she doesn't contaminate the scene with her DNA and to protect herself from biohazards – blood, vomit, faeces and the like.

She then walks through the scene, putting stepping plates down on the floor to protect the surface. On her first pass through she will look for evidence that may help identify offenders quickly. She will fast-track this kind of 'inceptive' evidence. For example, a fingerprint in blood on a window that the offender has climbed out of, or drips of blood going out of the house and down the street. In the UK, it's possible to get a DNA profile in just nine hours on a straightforward bloodstain, and the cost goes down depending on the turnaround time required.

In the US, it varies depending on the jurisdiction, but even top-priority jobs can take a few days and some can take months. Despite the beefing up of laboratory equipment and personnel, the increasing demand for DNA analysis has grown even quicker, and law enforcement is faced with a backlog of cases. As Peter observes, 'We have to think about what is needed to get us the results we need. Some of the things you see routinely on the TV happen once in a blue moon. They are generally a last resort. But timings are important in legal terms. Forensic teams need to sleep in order to perform properly. But once the police arrest a suspect, their custody clock is ticking and it's our job to provide results from inceptive evidence that inform their decisions about charging someone. There's always a balancing act going on.'

While these management decisions are being considered, the work at the scene continues. CSIs stand in each corner of a room and take a photograph of the opposite corner. They cover every aspect of every room, including the floor and the ceiling, so that if they move a piece of evidence they can see where it came from. Sometimes nothing will seem relevant, then, ten years down the line, a cold case review team will spot something crucial.

CSIs can also put a rotating camera in the middle of a room. It takes a series of photos which software then stitches together, enabling a jury to visualise walking around the room, and looking at particular objects. They can even click on a door and go through to the next room. 'For example,' says Peter, 'if several bullets have been fired through a window and have gone through walls and hit someone inside a house, you can scan the room and at a later date you can step back virtually from the house and show the trajectory of the bullets very, very accurately – to the point where you can show where the shooter was standing.' In this way, two key scenes of actions – the street outside and the place of impact inside – can be linked for the benefit of the jury.

Similarly, right from the start that afternoon in Bradford, CSIs were dealing with the street where the gunmen struck, and the interior of the travel agency, where staff had been threatened, pistol-whipped and tied up. In the street, there were bloodstains that had to be photographed and analysed by blood spatter specialists, to corroborate witness accounts of what had happened and in what order. A fingertip search revealed three bullet casings from a 9 mm handgun – one of the standard firearms used by career criminals, easily obtained even in the UK where guns are illegal.

Inside the travel agency, a careful search produced a handful of key pieces of evidence: a laptop bag that had been used to conceal the guns; a knife wielded by one of the men; and a bullet embedded in the wall. Ballistics experts identified the type of gun the bullet had been fired from. Nowadays, gun barrels have spiralled grooves – 'rifling' – running down their inside, which encourages bullets to spin, and hence fly more accurately. Each gun model has slightly different rifling. Back in Bradford, having examined the nicks and scratches on the bullet from the wall of the travel agency, the ballistics experts were able to determine that the bullet had been fired from a MAC-10. Later they explained that the MAC-10 had probably jammed, which might well have saved lives that afternoon.

Though the experts in Bradford would have been using powerful microscopes and vast digital databases in their identification, as a branch of forensics, ballistics has its roots in nineteenth-century detective work. Back then bullets were made in individual moulds, often by the owner of the gun, rather than mass-produced in factories. In 1835, Henry Goddard, a member of the Bow Street Runners (the first detective force in the UK), was called to the house of a Mrs Maxwell in Southampton. Joseph Randall, her butler, claimed that he'd been in a shootout with a burglar. He said he'd fought him off at the risk of his life. Goddard noted that the back door had been forced, and that the house was in disarray, but he was suspicious all the same. He took Randall's gun, ammunition, moulds, and the bullet that had been fired at him, and discovered that they all matched: the bullet had a tiny round bump that corresponded to a similar-sized flaw in Randall's mould. Confronted by this evidence, Randall confessed that he had staged the whole event in the hope of getting a reward from Mrs Maxwell for his bravery. This was the first time a bullet had been forensically traced back to a particular gun.

The scene may be the silent witness to the crime, but there are often human witnesses whose evidence can provide leads. In the Sharon Beshenivsky case, witnesses revealed that the robbers had made their getaway in a silver 4x4 SUV. At once, the traffic police began scanning local CCTV footage. They soon spotted the vehicle and identified it as a Toyota RAV4. A few months earlier, that could have been the end of the story. But earlier in 2005, Bradford had become one of the first cities in the UK to install a ring of cameras that recorded every vehicle entering and leaving the city. Up to 100,000 images a day were taken and stored by the Big Fish programme.

The police lost track of the vehicle when it left Bradford city centre. But when its number plate was put into the national Automatic Number Plate Recognition system (ANPR), analysts were able to tell detectives that the silver 4x4 had been hired at Heathrow airport. Within hours, the Metropolitan Police had found the getaway car and arrested six suspects.

But again, the Bradford detectives' luck seemed to have run out. The six men who were arrested were quickly able to prove that they'd had no part in the fatal robbery in Bradford. They were released without charge. It looked like the police had hit a dead end.

Once more, the CSIs came to the rescue. Once processed, a search of the RAV4 revealed rich sources of evidence: a Ribena drink carton, a water bottle, a sandwich pack and a till receipt. The receipt came from the Woolley Edge service area on the M1 south of Leeds. It was timed at 6 p.m., barely two hours after the fatal encounter between the gunmen and Sharon Beshenivsky. All these items were classic examples of inceptive evidence that can be fast-tracked for speedy identification.

When police examined the CCTV footage from the shop at the service station they spotted a man buying the items found in the RAV4. Meanwhile, those items were being examined for fingerprints and DNA and, when the results were run through national databases, the police found the names of six suspects, all connected to a violent criminal gang in London.

Now, it was only a matter of time before the guilty men were tracked down. Three of them, who had acted as drivers and lookouts during the robbery, were convicted of robbery and manslaughter. Two were convicted of murder, and given life sentences. One had managed to flee the country to his native Somalia dressed in a burkha, pretending to be a woman. But West Yorkshire Police refused to give up and, after an undercover Home Office extradition, finally he too stood trial and was sentenced to life in prison. Sharon Beshenivsky's law enforcement family never laid down. They made sure that every available resource was thrown into the hunt for justice.

CSI teams don't just pull out the stops for the headline cases. In volume crime cases, like burglary, if there is a realistic chance of recovering forensic evidence and identifying an offender they consider DNA swabbing, fingerprinting, footprint analysis. Sometimes they get an answer from one test that means they don't have to do any more complicated tests on that item. If they find fingerprints on a knife that's been used in a stabbing, they don't need to look for DNA on it. Peter explains, 'We don't want to be doing the really space age stuff when we can get the result we need by simpler and cheaper means.' This principle is occasionally passed over by some officers who enthusiastically watch TV crime dramas. Forensic scientist Val Tomlinson says, 'It can be your Senior Investigating Officer who's not had a lot of time on the ground. I remember going to one scene where a man was sitting dead with a knife in him and the SIO said, "Oh, so you will be doing metal analysis on the edges of the cut so that you can show it was that knife?" And I said, "Maybe that's not our priority given the knife is sticking out of him."'

But when the space age stuff needs to be done it can be done, as many cases in this book will show. Peter is particularly fond of the UK National Footwear Database, which can link scenes together through footprints. He used it recently after finding a rare footprint at the scene of a sexual assault. The print had been found at a few other crime scenes throughout West Yorkshire, and the coincidence made the police focus their attention on the man who was eventually convicted.

For Peter, it is the successful cases that stick longest in the mind. 'I remember one where we had a really good result, which isn't something you get every day. One of our CSIs went to photograph a woman who had been badly beaten up and was in intensive care. The woman subsequently died from her injuries but when the CSI saw her, he noted some odd shapes on her face. So we sent out one of our specialist imaging officers who took more shots using ultraviolet and infrared. When the photos were examined, we could see they were clear impressions of trainer soles.

'Later, when we recovered the suspect's shoes, not only did we find blood on them but our footwear expert could state that the victim had been stamped on at least eight times, because he could point out the same pattern in at least eight different orientations. His evidence made it clear that she had sustained a drawn-out attack. The suspect claimed he might have "accidentally stepped on her face". But I think the reason he got handed down a high sentence in court was the unequivocal forensic evidence.'

At the end of the long process of crime scene investigation lies the courtroom, where the evidence Peter and his colleagues collect is tested to its limits by lawyers and weighed in the balance by judges and juries. It's about as far from the dispassionate world of the scientist as it's possible to imagine. And it is no respecter of persons, as Peter recalls.

'I do remember being cross-examined in the witness box for about three hours on one occasion. There was clear DNA evidence that implicated the suspect for a violent robbery of a woman. But I would have to say I'd gone to unusual lengths to locate and recover this evidence. Perhaps above and beyond what you would normally expect.

'There could be no challenge on the DNA itself, but the defence took the line that I had planted this evidence. My integrity was what had to really withstand the test, and that was where documentation became really important. I was able to produce the original photographs taken before I touched or moved anything, so the jury could see the pristine scene. The photographs were taken sequentially as I recovered the items and eventually we got to the item that produced the DNA profile. The jury could see exactly what I did, what order I did it in, the unique identifiers on the item.

'Then I was challenged over whether someone had tampered with it afterwards. But I was able to document every process it had gone through. There was a clear chain of continuity. Still the attack went on. I ended up putting on a scene suit, a mask, gloves and a hairnet and placing a sterile piece of paper out in the courtroom. Then I opened up the exhibit. I showed it to the jury then showed my photographs to show that it was the exact same exhibit with the exact same unique identifiers. The evidence stood up to the test, but it showed me the lengths that a defence will go to to try and get the client off.

'I found that quite annoying personally, but I actually see the need for an adversarial system. I was challenged but ultimately that strengthened the case because it was clear that there were no issues with the evidence. Ten years down the line we're not going to have an appeal in that case saying the evidence could have been tampered with. I'd rather get it out in the open now. Let's challenge it now, let's face the scrutiny.'

Technology has come a long way. But there is still a distance to go. And we creators of fictitious murder don't always help, as Peter confirms. 'The public's expectations are often raised by what they see on TV. When we turn up and explain why we can't examine something, sometimes they just don't believe us. We end up feeling like the bad guy because we can't deliver what they're expecting.'

Peter is referring to the 'CSI effect', named after the famous TV series *CSI: Crime Scene Investigation*, which some say has distorted the public perception of what forensic science is capable of doing. DNA evidence, in particular, has come to be seen by quite a few jurors as indispensable evidence. However, the extent of the CSI effect is disputed by many, who instead see it as a sign that ordinary people are getting a basic grip of what forensic science is doing, even if it is an imperfect one. When experts and judges do their job properly, they can help juries understand the significance of different types of evidence.

In one extraordinary case in Wiltshire in 2011, a crime victim copied a trick she'd seen on an episode of *CSI*, to help the forensics team she hoped would investigate. For months, a man had been prowling around Chippenham in his car. He would identify a woman, pull on a black balaclava and gloves, and drag her into his car. Then he would drive off, sometimes to a disused barracks, where he would rape her, and then force her to clean herself with towels to destroy the forensic evidence. He was caught out when his final victim tore out some strands of her own hair and left them in his car, before he let her go. She told the police that, whether or not she survived, she knew there would be an investigation, and that this would provide DNA evidence. 'I've always been a fan of *CSI* programmes. I've watched so many of them, I know what to do and how things work.' With the help of her hair, and the spit that she'd also deposited on the car seat, Lance-Corporal Jonathan Haynes was convicted of six rapes.

In certain ways Peter Arnold thinks that British CSIs should be more like their TV counterparts. 'We really need a decent mobile data solution for CSIs which will enable them to have proper IT access at the scene to process information and record exhibits so they don't have to keep coming back to their base, which wastes a lot of time. That sounds really easy, doesn't it? Because I can walk around with my iPhone and I can have all sorts of things available at my fingertips. But there's the costs of developing and providing

Detectives combing the area around Sharon Beshenivsky's murder scene for evidence

the software. We haven't got millions of pounds to develop an app for CSIs. And then there's the issue of data security.

'But if we can develop real-time forensics, what a difference that could make. If somebody's house gets broken into and we find some potential DNA evidence, we've still got to get that DNA evidence from the crime scene via a courier to the lab. It's got to be booked in and finally it's got to be processed. Currently we are fast-tracking certain evidence from burglary scenes, turning DNA around within nine hours because burglary is such a priority. Why wait two or three days to get a burglar's ID when you can have it in nine hours, get them in custody and stop them burgling someone else tonight? So we're using major crime principles for volume crime. Similarly with fingerprints. We've really speeded that up, but if we could scan the print at the scene that would speed it up even more.

'Imagine this. If we get to a burglary within an hour and we examine it within half an hour, we could potentially have a burglar's name within an hour and a half of the crime being discovered. The police then go and knock on their door and they're still there with the stolen property in the bag. So the victims get their stuff back. And the burglars start to understand there's no point.'

As well as the satisfactions, the job comes with stresses and pressures. We make high demands on the people we expect to deliver justice, and we don't always appreciate how much it eats away at them. Peter Arnold says, 'We see some of the worst things that mankind can do to each other and I still get shocked by some of the things that occur. Most people can go home and talk to their families about what they've done at work. We can't. But even if I could, I don't want my family to know some of the things that I've seen.'

TWO

FIRE SCENE INVESTIGATION

'It's usually pretty dark, smelly, uncomfortable and physically demanding. The days are long and you come home filthy and stinking of burnt plastic. There's nothing glamorous about it. But it is fascinating.'

Niamh Nic Daéid, fire scene investigator

Sunday, 2 September 1666. A family servant coughs himself awake in Pudding Lane, London. Realising there's a fire in the shop below, he pounds on the bedroom door of his master, baker Thomas Farriner. The whole household crawl along rooftops to safety, except for the maidservant Rose, who, paralysed with fear, perishes in the blaze.

Soon flames begin licking the walls of neighbouring houses and the Lord Mayor, Sir Thomas Bloodworth, is called upon to authorise the firemen to pull buildings down to stop the fire spreading. Bloodworth is angry at having his sleep disturbed and ignores the firemen's urgent demands for drastic action. 'Pish!' he says. 'A woman could piss it out.' And leaves the scene.

In the middle of the morning, diarist Samuel Pepys experiences 'the wind mighty high and driving [the fire] into the city, and everything, after so long a drought, proving combustible, even the very stones of churches'. By the afternoon London is in the grips of a firestorm, roaring through 'warehouses of oyle and wines and Brandy', wooden houses, thatched roofs, pitch, fabrics, fats, coal, gun powder – all the flammable material of life in the seventeenth century. The immense heat of the blaze makes escaping gases rapidly expand and rise, sucking in fresh air at gale force speeds, feeding the inferno with yet more oxygen. The Great Fire has created its own weather system.

When the fire abates four days later, it has destroyed most of the medieval City of London, including more than 13,000 houses, 87 churches and St Paul's Cathedral. Roughly 70,000 of the city's population of 80,000 are suddenly homeless.

The ashes were still warm when the conspiracy theories started. Most Londoners could not bring themselves to believe the fire was an accident. There were too many coincidences for that; it started among tightly packed wooden buildings, while everyone slept, on the one day of the week when the streets were empty of helping hands, when a gale was blowing, and the Thames lay at low tide.

Rumours of foul play were rife. A surgeon, Thomas Middleton, had stood at the top of a church steeple and watched as fires seemed to break out in several distinct and distant places at once. 'These and such like observations begat in me a persuasion that the fire was maintained by design,' he wrote.

Foreigners in particular were suspected, with one Frenchman beaten nearly to death in Moorfields for carrying 'balls of fire' in a box; they turned out to be tennis balls. Poems and songs expressed the bewilderment at the fire's origin and cause:

It's still unknown from whence our ruine came;
Whether from Hell, France, Rome, or Amsterdam.

Anon., 'A Poem on the Burning of London' (1667)

The desire to know the truth started at the very top. Charles II had lost more property in the fire than anyone else. The king empowered parliament to set up a committee of inquiry into the fire. Scores of eyewitnesses came forward. Several said they had seen people throwing fireballs, or confessed to throwing them themselves. One Edward Taylor said that on Saturday night he went with his Dutch uncle to Pudding Lane, found the window of Thomas Farriner's bakery open, and threw in 'two fireballs made of gunpowder and brimstone'. But as Edward Taylor was only ten years old his account was dismissed. Robert Hubert, the simple-minded son of a French watchmaker, confessed to having started the fire. No one really believed him but because he insisted the jury found him guilty, and he went to the gallows at Tyburn.

One member of the parliamentary committee, Sir Thomas Osborne, wrote that 'all the allegations are very frivolous, and people are generally satisfied that the fire was accidental'. In the end, the committee decided that the dreadful conflagration was caused by 'the hand of God, a great Wind, and a very dry season'.

It's not surprising that the committee arrived at so unsatisfactory a conclusion. For investigators to evaluate complex fire scenes, they need to understand how fire works. Back in the seventeenth century, the scientific knowledge was woefully insufficient. It wasn't until 1861, when Michael Faraday put his lectures on fire into a book, that such understanding became readily available to a wide audience. *The Chemical History of a Candle* was the published version of six lectures he delivered for a young audience, but it is still regarded as a key text on the subject. Faraday used the candle as a symbol to illuminate the general nature of combustion. In one key lecture he snuffed a candle out by putting a jar over it. 'Air is absolutely necessary for combustion,' he explained. 'And, what is more, I must have you understand that fresh air is necessary.' By 'fresh air' he meant 'oxygen'.

Faraday was an early expert witness, taking with him his findings from the laboratory – sometimes quite literally. In 1819 the owners of a sugarhouse destroyed by fire in Whitechapel, London, sued their insurance company who had refused to pay out £15,000 compensation. The case turned on whether or not a newly developed process involving heated whale oil – which the owners had started using at the factory without the insurers' knowledge – had made the fire more or less likely. Before

Michael Faraday, whose 1861 book The Chemical History of a Candle *paved the way for modern fire scene investigators*

testifying, Faraday performed experiments on whale oil, heating it to 200°C to demonstrate that 'all the vapours of the oil, except water, are more inflammable than the oil itself'. In court a member of the jury did not believe him, so Faraday set fire to some of the oil's distilled vapours (naptha) which he had brought with him in a vial, 'a most offensive smell being at the same time perceived throughout the court'.

Faraday's most important forensic investigation was into an explosion at Haswell Colliery, County Durham, which killed ninety-five men and boys in 1844. The blast occurred at a time of industrial unrest in the Durham coalfield. The lawyer acting for the grieving relatives petitioned the Prime Minister, Robert Peel, to send government representatives to the inquest. Faraday was among those sent.

The team spent a day visiting the mine, investigating in particular its air flows. At one point, Faraday realised he was sitting on a keg of gunpowder near a naked candle flame. He leapt to his feet and 'expostulated with them for their carelessness'. The jury reached a verdict of accidental death, with which Faraday agreed. But the team submitted a report on their return to London noting that coal dust had played a major part in the explosion, and recommending that the ventilation be improved. The mine owners objected, because of the costs of improvement. The risk was ignored for sixty years, until a similar explosion in 1913 led to the death of 440 miners at Senghenydd Colliery in Wales – the worst mining disaster in UK history.

In the twentieth century, the Fire Service and the scientific community developed fire scene investigation in tandem, encouraged by governments who wanted to know how many fires there were, their origins and their causes. In the 1960s and 1970s investigations became more rigorously scientific: protocols were adopted; new instruments enabled complex chemical mixtures such as gasoline to be identified at fire scenes; and experts in the field began to emerge. Partly as a result of this increased understanding, it is now rare for a fire or an explosion – which is essentially an expedited fire – to cause such horrendous loss of life in peacetime. But when they do, they leave an indelible impression on those who investigate them.

Among those who became the new experts in fire investigation were an Irish husband and wife. Their daughter, forensic chemist Niamh Nic Daéid of the University of Dundee, has continued their legacy, searching for the truth amid scenes of terrible destruction. Niamh explains. 'I have a legacy in forensic sciences, if you like, because my parents were both independent fire investigators, and indeed my mother still does fire scene investigation, so I grew up with it. Myself and my brother used to make our pocket money by sticking Mum and Dad's fire pictures into reports – for five pence a picture. As you can imagine, the conversation around the dinner table was always about fires.'

Whether fire obliterates someone's property or their dearest relative, the investigator works at the point of collision between nature's most violent force and the human world it wrecks. I was forcibly reminded of that when I asked Niamh about fires that had particularly affected her. The first words out of her mouth were: 'The Stardust disco fire.'

I was asleep in my bed in Derbyshire in the early hours of Valentine's Day, 1981. I was a young journalist, based in the northern newsroom of a national Sunday paper. I'd never covered a major disaster but I knew that was about to change when a ringing phone woke me in the small hours of the morning. The familiar gruff voice of my news editor said, 'There's been a major fatal fire in a Dublin disco. It's looking like dozens dead. You're on the seven o'clock plane.'

By the time I got to Manchester Airport, the radio had confirmed what I'd already been told. A massive fire. A horrifying death toll of young people who'd set off for a fun night out and who wouldn't be going home. Inside the airport, journalists and photographers milled around, looking for colleagues so they could hunker down in their own little scrums and divide up the tasks on the ground at the other end.

My own team – three other reporters and two pic men – made our way to a corner of the bar. A double whisky was set in front of me. Even in those hard-drinking newspaper days, I wasn't accustomed to starting my day that way. 'Drink it,' one of my colleagues insisted. 'Trust me, you're going to need it before today's over.'

He was right. When we landed at Dublin, our Irish staff reporter gave us the grim news. More than forty dead. Because I'm a woman, because I was considered to be good with the grieving but also hard-headed enough to get what I'd come for, I was assigned to the death knocks – visiting the bereft families so we could flesh out our story with poignant quotes and photographs of the dead.

I spent the rest of the day on the Coolock council estate, where many of the teenagers who died in the Stardust had come from. The families were in shock, but oddly grateful that someone wanted to mark the passing of their children. I'd never spent a more harrowing day at work. And I was just a spectator. My heart felt hollow when I imagined what the bereaved were going through.

After the first edition deadline had passed, I met up with one of my team at the site of the fire. From the front of the building there wasn't much to see except broken windows and smoke staining the upper part of the facade. Apart from the throat-catching stink of smoke and char, it was hard to believe forty-eight people had died and more than 240 had been injured there. It was the interior of the building that had been devastated by the fire; from the outside, the only giveaway was the number of fire engines and police vehicles crowding the roadway outside.

Niamh Nic Daéid's mother was one of those charged with finding out what happened inside the Stardust disco that night.

The Valentine's Dance at the Stardust was supposed to be a night to remember for very different reasons. Eight hundred and forty-one people, mainly in their late teens, had handed over the £3 entrance fee that entitled them to sausage and chips and the right to dance till two in the morning, thanks to a special late licence.

Twenty minutes before closing, the DJ announced the winners of the best dancing prizes. A minute later, some revellers spotted smoke coming from behind a roller blind to the left of the dance floor. Most of them put it down to a special disco effect and kept dancing.

Behind the blind were five rows of tiered cinema seats. When some of the dancers looked behind the blind they saw some of the seats on the back row ablaze. Their polyurethane stuffing was already emitting black clouds of extremely poisonous hydrogen cyanide. At first the fire was small and controllable, but it quickly grew in intensity. Employees emptied water extinguishers into the flames – to no avail. Within five minutes, molten plastic was dripping on to the patrons on the dance floor; part of the ceiling collapsed on to them; and thick, toxic smoke filled the entire ballroom. Survivors spoke of their shock at the swiftness of it all.

Fire scene investigators at the scene of the Stardust Disco Fire, in which forty-eight people died and more than 240 were injured

When people panic they instinctively try to leave a building the same way that they came in, so the narrow foyer leading to the Stardust's main entrance quickly became a bottleneck. Those sprinting to the main doors found them locked shut and it took a bouncer crucial minutes to squeeze through the desperate crowd with the key.

But still disaster should have been averted. There were six fire exits in the Stardust. But the owner, Eamon Butterly, had been worried about people opening the doors from the outside and slipping into the venue without paying, so one of the fire exits was locked and others had chains wrapped around them so they appeared locked. Panic-stricken patrons tried, and eventually managed, to kick these doors open. Another fire exit had tables and seats stacked on either side of it; yet another had plastic skips blocking it.

At 1.45 a.m., when the ceiling in the ballroom collapsed and the electricity cut out, around 500 people were still inside. The blistering flames were their only source of light. The Adam and the Ants record that had been playing was replaced by terrified screams. Within nine minutes of the fire being spotted, everything in the Stardust was ablaze – seats, walls, ceiling, floor, tables, even metal ashtrays.

In the mayhem, some people fled into the toilets. Six weeks before the disco, Butterly had heard customers were trying to smuggle alcohol in through the toilet windows, so he'd had steel plates welded to their inside, to complement the metal bars that were already in place outside. When fireman arrived at the scene, eleven minutes after the fire started, they attached cables to the bars and drove away at speed, but only managed to bend them. The people in the toilets were trapped in an inferno of flame and smoke.

Everyone in the surrounding area, in the working-class communities of Artane, Kilmore and Coolock, knew someone affected by the tragedy. The whole of Ireland mourned the forty-eight killed. Five of the dead had been so badly burnt that they could not be identified. (In 2007 their bodies would be exhumed from a communal grave, so DNA analysis could separate and identify them.)

At 8.35 on the morning of Valentine's Day, Detective Garda Seamus Quinn inspected the gutted Stardust. He spent five hours examining the site, finding no trace of accelerants or electrical problems in the area where the fire had first been spotted. He also discovered, by throwing a lit cigarette on to a similar seat, that its non-flammable PVC covering did not catch fire. Had someone slashed a seat and deliberately lit its polyurethane filling?

The British Fire Research Station carried out a full-scale reconstruction of the area where the fire had first been spotted at their hangar at Cardington, Bedfordshire. Investigator Bill Malhotra managed to get seats to catch light both by slashing them open, and by placing several sheets of newspaper underneath them. The flames reached the very low ceiling and started melting the carpet tiles, causing molten drops to fall on to other seats. Within the tight space all the seats heated up, and the boiling drops were enough to defeat their PVC coverings. Once five seats in the back row were in flames, the row of seats in front caught fire, too. Quinn's and Malhotra's experiments both suggested arson.

In June 1982, eighteen months after the blaze, the Irish government published the results of a public inquiry into its origin and cause. On the question of why, the report was ambiguous. 'The fire was probably caused deliberately,' it said at one point. At another, 'The cause of the fire is not known and may never be known. There is no evidence of an accidental origin and equally no evidence that the fire was started deliberately.' The forensic experts who had given evidence were divided. While Quinn, Malhotra and one other thought the fire was most probably caused by arson, two others wouldn't rule out an electrical fault.

The report damned Eamon Butterly for many things, including not complying with electrical safety standards and using locks rather than doormen to guard doors. The cost of employing extra doormen would have been £50 – just over £1 for every life lost. On the question of the armoured toilet windows the report said, 'While their primary purpose was for ventilation, it might have been possible for a person to get through them in an emergency.' Despite all of these points, the report legally exonerated Butterly from responsibility for the fire because it had been 'probably caused by arson'. So in 1983 the state paid Butterly compensation for malicious damages in the region of £500,000. In 1985 the victims' families received an average of only £12,000 each.

The families were far less interested in money than in why their relatives had perished. So much potential evidence had been obliterated that it seemed unlikely they would ever get an answer. But that didn't stop them trying. In 2006 the Stardust Victims' Committee enlisted the help of a new set of forensic experts in a bid to make the case for a new public inquiry. Those experts pointed out that in the reconstruction in the hangar in Cardington the fire had taken thirteen minutes to burn through all the seats and had never breached the roof, whereas the real thing had shot from the first seat it was seen on – at 1.41 a.m. – up into the night sky within five minutes. Something didn't add up.

The experts also drew attention to various witness accounts that supported this view. Eyewitnesses standing outside the building said they had seen flames coming from the roof several minutes before 1.41 a.m. In the weeks leading up to Valentine's Day, Stardust employees had seen a smoke-like substance and 'sparks' coming from the Lamp Room above the Main Bar, which was right by the rows of blazing seats. On Valentine's Day itself, Linda Bishop and her friend had been sitting below a grille in the ceiling, listening to 'Born to be Alive', when they felt a great increase in temperature. Linda looked down at the new digital watch she'd been given for Christmas. It read '1.33'. A barman who had fought the fire on Valentine's Day said he had 'felt a monstrous heat coming from the ceiling. I was positive that the fire started up in the ceiling.'

The Stardust Victims' Committee experts came to the conclusion that the burning ceiling had set fire to the seats, rather than the other way around. They believed that an electrical fault in the Lamp Room – which was located in the roof space and contained spot-lamps and plastic seats – had ignited the ceiling. Right by the Lamp Room was a storeroom, and the experts thought the original inquiry had been misled about some of its contents. Eamon Butterly's solicitor had provided a list of the 'approximate contents' of the storeroom, including 'bleaches, brio wax, aerosols, gasoline-based waxes and polishes', but did not mention the highly flammable 'drums of cooking oil' which were also present.

Professor of Fire Dynamics Michael Delichatsios reasoned that if enough heat had come from the Lamp Room, the highly flammable contents of the storeroom would have spontaneously combusted. This would account for the extreme speed at which the fire spread, raining burning plastic on to the heads of people on the dance floor, and eventually bringing the whole ceiling down. In 2009 the government commissioned Senior Counsel Paul Coffey to examine the Stardust Victims' Committee case for a new public inquiry. He found the 'probably deliberate' finding of the original inquiry 'so phrased as may well give the mistaken impression ... that it is established by evidence that the fire was started deliberately and not a mere hypothetical explanation for the probable cause of the fire'. He recommended against a new inquiry, but suggested the government change the public record to make it clear that the cause of the fire is unknown. So, twenty-seven years after the most lethal fire in Ireland's history, the government officially clarified the cause of the fire as unclear. Because the Lamp Room had been 'totally destroyed', 800 eyewitnesses and scores of dispassionate forensic scientists could never know if it was the true origin of the blaze. The secrets of the fire were destroyed with it. And that is often one of the frustrations of fire investigation.

Fire scenes vary in their complexity, but even relatively simple ones challenge the investigator, who must try to reconstruct a destructive chain of events. Let's take a typical scenario. A passer-by sees a house on fire and calls the fire brigade, who put it out. A structural engineer declares the building safe to enter, and a fire scene investigator like Niamh Nic Daéid arrives to determine the origin of the blaze, why it happened, and how it spread.

First of all – and unusually for a forensic practitioner – Niamh may sometimes interview eyewitnesses. Where exactly did they see the fire? Was there yellow flame and white smoke, which gasoline gives off, or the thick black smoke of burning rubber? Getting the best out of eyewitnesses is a skill. Niamh is often talking to people on the very edge, sometimes after the centre of their world has burnt down. Occasionally the fire investigators may have to 'stop the interview and let the police know that this person might have turned into a suspect'. It is a well-known axiom that industrial fires increase when business conditions get tough, as some firms consider the advantage of a successful insurance claim over a loss-making factory. Arson aside, when accidents do happen people can be cagey. When Niamh asks employees where they were smoking before a fire in their office started, they usually say in the allocated area. But experience has told her that 'when it rains people tend to smoke beside the back door, where the rubbish is'.

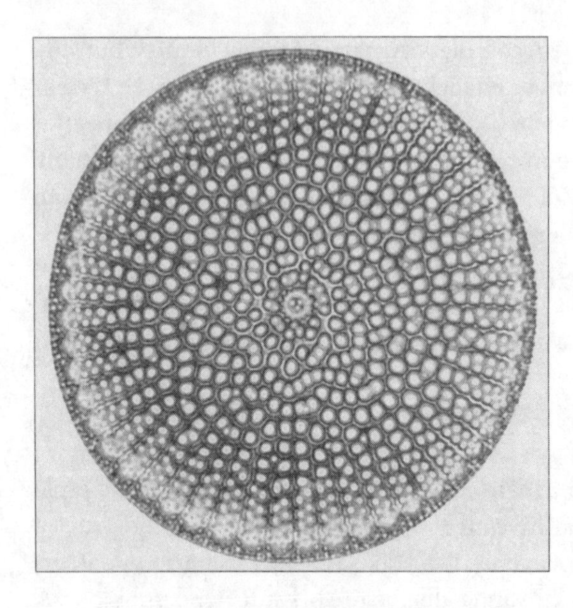

The fossilised remains of a diatom – a single-celled organism – viewed under a microscope

The talking over, Niamh walks around the outside of the building and lets things sink in. Are there patterns of smoke on the walls? Which windows are broken? Anything potentially significant on the grounds, like a gas can or cigarette ends scattered about? Then she walks through the building 'hands in pockets, not picking things up', looking for anything unusual. Now she's ready to get dirty. Outside she deals with the gas cans and cigarette ends she saw earlier, 'photographing them *in situ* with a scale if possible, drawing them on a plan, packaging and labelling them appropriately'. Inside she approaches 'the business end' – where the fire most likely started – in a systematic way, moving from areas of least to most damage, documenting and photographing the scene as she progresses.

As a fire spreads from its point of origin it creates more heat, which ignites more material in a self-sustaining chain reaction ruled by the supply of fuel and oxygen. By the time it stops burning it has often brought down ceilings and walls, which shield things as they fall. The scene is all the more resistant to interpretation once firefighters have directed thousands of gallons of water into it. 'So you've got your burnt-out shell of a house with material all over it. In order to get to the bottom of where it started, you need to de-layer it, like an archaeological dig.' Like a pathologist sawing open a ribcage to perform an autopsy, Niamh has to cause more destruction to reach her answers. She works from the area of least damage inwards because 'if the big black hole in the corner is where the gas was poured, and you march over to that and walk around in it, you'll cross-contaminate your scene'. In extreme cases, the investigator will use tape to construct a grid on top of the scene, number each square and take everything out in buckets to sieve for any evidence that might have survived.

Because fires tend to rise and spread sideways, they sometimes leave the charred outline of a 'V' pointing at their origin. Things are less clear cut when an arsonist has sloshed gasoline throughout a house. Thin lines of severe burning on the ground, surrounded by milder burning, can indicate a gasoline trail, but flames follow the gasoline's path with such speed that a single point of origin is well nigh impossible to discern. If Niamh finds several widely distant instances of equally bad burning, this may also indicate arson; two unconnected accidental fires beginning at the same time in a house is a vanishingly rare occurrence.

Once Niamh has found the most likely origin/s of the fire, she looks for potential sources of ignition – matches, lighters, candles; and fuel – TVs, newspapers, rubbish bins. Arsonists often leave matches behind, assuming they will burn away to nothing. But the powdered rock in a match head contains the fossilised remains of single-cell algae called 'diatoms'. A diatom's shell is made of silica, which is abrasive enough to help you strike the match, and tough enough to endure extremely high temperatures. Each of the 8,000 known species of diatom has a unique shell structure, identifiable through a microscope. Different brands make their matches using powdered rock from different quarries. If forensic scientists can spot the diatoms, they can identify the match brand. Then a search of a suspect's pockets or CCTV footage from local shops can provide incriminating evidence.

In her mind Niamh tries to imagine how the scene was arranged when the fire broke out. Then, as far as possible, she reconstructs it for real. Fire investigators don't always get this right, as Niamh once experienced in the case of a suspicious house fire that had begun on a desk. The police asked the fire investigators to put the sooty items back on the desk in their original positions. When Niamh was called in to review the scene, she thought it best to do her own reconstruction and compare it to theirs.

'The other investigators had reconstructed it in a way not sustained by the physical evidence, not noticing things like a circle where a cup had protected the desk from smoke. They'd put the items in the wrong place and taken photographs which told an incorrect version of the story. With the items put back in their correct position the set of circumstances that created the fire came to life.' In 2012 Niamh ran a series of workshops relating to fire scene investigation in Scotland, which concluded that, whilst many investigators are very well equipped for the job, '97 per cent of fires in Scotland are investigated by personnel who have less than a week's training in fire scene investigation'. While many of these fires are relatively straightforward to investigate, the point relating to appropriate training still remains. Trained fire investigators are critical in the correct determination of the origin and cause of a fire, and this is particularly the case 'in fire fatalities where the investigators have a huge obligation both to the victims and to their relatives to be able to say how individuals died in that fire.'

Mishandling evidence leads to confusion and to conflicting versions of events being presented in court. It's crucial to get it right the first time, not least because the clues are often so fragile. Can you get fingerprints? Can you get DNA? Can you recover information from a hard drive in a melted computer? 'The answer to all of those is "yes", if you have the awareness not to go clumping around damaging material.'

Treading lightly is not easy for Niamh in her heavy-duty steel-toe-capped boots, hard hat and protective overall. The scenes that she enters can contain live electrical hazards, jagged glass, partially collapsed walls. 'It's usually pretty dark, smelly, uncomfortable and physically demanding. The days are long and you come home absolutely filthy and stinking of burnt plastic. There's nothing glamorous about it. But it is fascinating.'

At the suspected point of origin, Niamh collects the debris and sifts through it by hand. 'You would be astonished at what survives. Fires are destructive things, but they generally leave quite a lot of material behind. Things like buttons, lighters, bottles, beer cans, anything metal, survive relatively well. Plastic materials can be melted on one side but fine on another. So you might be able to lift a fingerprint from the underside of a TV remote control.'

Electricity can be your friend in the fire scene, and can provide corroborating physical evidence relating to the cause, origin or spread of the fire. Fire investigators like Niamh crawl around in the muck armed with pliers, following cables as if they were Ariadne's thread guiding her through the labyrinth. 'Many scene investigators don't see the value of the electrical circuitry. It's very laborious and time-consuming work, but enormously useful because it gives solid physical evidence, compared to burn patterns which can be interpreted more subjectively.'

On the wall of her office Niamh has two photographs of a 12-storey building by Piccadilly Tube station in London. The top seven storeys were wrecked by fire to the tune of £12 million pounds' worth of damage. When the investigators first arrived at that scene they spoke to a cleaner who reported that she'd spotted the fire when it was still small, in the lighting system of one particular floor. That gave the investigators a pointer, but finding the exact origin of such a severe fire was nevertheless still daunting. Niamh spent two days in the building with her colleagues before they finally tracked it down to an electrical fault within a water cooler. 'It was a really interesting fire because it involved a lot of use of the electrical system to corroborate the area of origin. So it's dear to my heart, which is why I've got a picture of it on the wall.'

Some fires begin with electrical faults. But others have less innocent origins. Fire scene investigators will often bring in sniffer dogs, whose sense of smell is 200 times more sensitive than a human's, to find ignitable liquid accelerants like gasoline, paraffin and paint thinner. There are about twenty hydrocarbon dog teams in the UK, many of whom wear little boots to protect their paws (and to protect the scene from contamination). 'I've seen them in action and, boy, are they good. They just sit down and indicate when they smell something,' says Niamh.

Once a dog has identified the presence of a hydrocarbon, the fire investigator starts to bag the evidence. Because plastic bags react with the hydrocarbons in substances like gasoline, they put suspicious material into nylon bags and takes it back to the forensic science lab for analysis. If the material is something like a piece of carpet, the investigator tries to take a separate, unburnt piece from the scene, for comparison. In the laboratory forensic chemists will analyse the fire debris submitted. They use various techniques to extract possible chemical accelerants, including 'headspace extraction'. The most common way of doing this involves placing the material in a closed container and heating it to allow vapours to rise off it. These are then collected using an absorbent material, and extracted using a chemical solvent. From this vapour the forensic chemist tries to identify particular compounds, usually using gas chromatography. This is a fairly complex scientific process which causes the chemical molecules within the vapour mixture to separate according to their size. Niamh explains: 'If you can imagine a drainpipe that's ten feet long and you pour molasses down it, so the inside is coated with molasses, and then you get a box of marbles of different sizes and you pour them down it, the little marbles will stick longer than the big marbles. So you get big marbles out first and then little marbles. That's what GC does, in a nutshell. Juries can visualise that, so they go, "Oh, now I get it."'

If the tests show gasoline, then, depending on the case, the next step may be to carry out 'gasoline branding'. Most molecules in a can of gasoline will evaporate at room temperature (which is why you can smell it), but manufacturers put additives in their gas which do not evaporate. The additives make car engines run more efficiently, and can survive very high temperatures. They are also quite specific to different brands. Additives are extremely stable and can stay on clothes until they are washed out with detergent.

Establishing a gasoline brand was important in obtaining convictions following one of the most distressing house fires in my recent memory. At 4 a.m. on 11 May 2012, a fire began burning the inside of the front door of 18 Victory Road, Allenton, Derby. Two minutes later, it had raced up the carpeted staircase to the open doorway of a bedroom full of sleeping children. Their father, Mick Philpott, called 999 – 'Help me! My babies are trapped inside the house!' Jade, John, Jack, Jesse and Jayden Philpott, aged between five and ten, died at the scene, and Duwayne Philpott, aged thirteen, died later in hospital, all from smoke inhalation.

Hours after the flames had been put out, Mat Lee from Derbyshire Fire Service arrived at the scene. A colleague had found an empty gas can and a glove near Victory Road, so Lee was on especially high alert for arson. He removed the top layer of debris from underneath the front door, and a hydrocarbon dog started barking. Lee packaged the material and sent it off to forensic chemist Rebecca Jewell for analysis.

Five days after the blaze, the parents of the dead children, Mick and Mairead Philpott, gave a press conference to thank friends and family for their support. But their behaviour aroused police suspicions. Assistant Chief Constable Steve Cotterill felt that Mick acted like an 'excited child' instead of a grief-stricken parent. 'I would have expected him to be completely and utterly destroyed,' Cotterill said later. 'It was a sham, in my view.'

The police put the Philpotts under 24-hour covert surveillance. A bug in the couple's hotel room picked up Mick telling his wife: 'You make sure you stick to your story,' and later, 'They're not gonna find any evidence, are they? You know what I mean?' On 29 May, the Philpotts were arrested on a murder charge (which was later downgraded to manslaughter).

Over a period of six months Rebecca Jewell received various samples from the scene and from the defendants' clothes. In the abandoned plastic tank, she found a mixture of additives including those from Shell gas. She found traces of gasoline in the carpet under the door of the house, but couldn't tell which brand it was because the additives were contaminated by a chemical from the carpet underlay. She found Shell additive on Mick's boxer shorts and right trainer. She found Total (another brand of gasoline) additive on leggings, a thong and a sandal belonging to Mairead, and on the clothing of Paul Mosley, who had been charged with helping the Philpotts set the fire.

When the trial began in February 2013 the jury was told that the Philpotts and Mosley had started the fire in a bid to incriminate Lisa Willis, Mick Philpott's former mistress. Lisa had spent ten years living in the house with Mick, their four children, her fifth child from a previous relationship, and Mairead and her children, but had recently left the house and taken her children with her to live with her sister. A custody hearing had been scheduled for the morning after the fire, and Mick Philpott had hoped to pin the arson on Lisa, to prevent her winning the right to keep their children. Mick and Mairead had put all the children to bed in one bedroom, and rested a ladder up against the bedroom window. The plan was for Mick to climb up and rescue them, making him look like both a victim and a hero. But the fire spread too quickly. There was no time to get in through the window and save the children. All three defendants were found guilty of manslaughter; Mairead and Mosley were sentenced to seventeen years in prison, and Mick to life. The Philpott fire dominated the media for weeks; the *Daily Mail* headlined an article 'Mick Philpott: Vile Product of Welfare UK'. While some were wondering if the Philpotts had been using the kids to generate their £13 per week child benefits, Niamh Nic Daéid's thoughts were somewhere entirely different. 'Why didn't the smoke detectors wake the kids?'

One of her Masters students had been part of the team investigating the fire. Together they decided that for his dissertation he would look into the ability of smoke alarms to wake children. They asked the parents of thirty children to set off the smoke alarms in their properties at random hours of the night. 'Eighty per cent of these children did not wake up, even though some of them had the alarm in their bedroom.' The variable frequency detectors designed to address the problem of heavily sleeping children seldom worked. Some of the most effective alarms are reported to be the ones that allowed the mother to record a message herself: 'So she says, "Get up!" and children respond to the pitch and frequency of her voice.' Now the challenge is to learn the lesson of the fire investigators' research – a challenge Niamh's research team is taking up with smoke detector manufacturers.

The desire for custody of children is possibly a unique instigation for arson. Much more common motivations include revenge, insurance fraud or the desire to cover up after a burglary, or even a murder. But people who try to dispose of a body by setting fire to a house or, like Jane Longhurst's murderer, torching the body itself (see p. 211), are unlikely to succeed. Any forensic investigator dealing with a fire quickly learns to distinguish between the normal effects of fire on a body and evidence which may have a more sinister explanation. Whether or not someone was still alive when the fire started, heat causes the muscles of the body to seize up, drawing the legs and arms up into a classic 'pugilist' stance. Water loss shortens the limbs and causes the body to lose up to 60 per cent of its weight. The facial muscles are distorted and the skin of the limbs and torso bursts, creating tears which an inexperienced investigator could mistake for wounds received prior to death. The bones, made brittle by exposure to heat, often fracture when the body is moved from the scene to the morgue. But, even if a body is badly charred on the outside, it will usually be surprisingly well preserved internally. At a crematorium bodies are reduced to ashes by exposure to 815°C heat for around two hours. While structural fires can reach 1,100°C, they generally don't stay hot enough for long enough to completely destroy evidence of foul play.

Some people love fires so much that they start them with no obvious motive at all. This is the pure arsonist. Their addiction starts small but invariably escalates, and is rarely overcome. It often incorporates a sexual element and can be fiercely compulsive.

One extraordinary serial arsonist started fires in Californian buildings in 1984 and didn't stop until he was arrested in 1991. During those seven years federal agents estimated he set more than 2,000 fires. Joseph Wambaugh wrote a book about him, *Fire Lover* (2002), and HBO made a feature film, *Point of Origin* (2002).

The story begins in 1987 when Captain Marvin Casey of the Bakersfield Fire Department was summoned to a fire in a fabric store. As soon as he got there he was called to another Bakersfield fire, this time in an arts and craft store. This second one had been extinguished before it overwhelmed the building, and Casey was able to recover a time-delay incendiary device – a lit cigarette placed alongside three matches, rolled up in a yellow sheet of notepaper, and held together with an elastic band. The arsonist had moved the cigarette up so its base was in contact with the match heads, giving him up to fifteen minutes before the cigarette burned down, and fire erupted.

Over the next few hours Casey heard of two more fires in Fresno, 100 miles down Highway 99 from Bakersfield. It felt like too much of a coincidence; Casey suspected that a serial arsonist was in play. Curiously, Fresno had been hosting an arson investigators' conference which ended shortly before the fires broke out.

Casey sent the incendiary device from the Bakersfield craft store to a fingerprint examiner, who managed to lift a good left ring finger off the yellow notepaper. He put the print through both the state and national criminal record databases, but found no matches.

Casey began to think the unthinkable. Could one of the fire investigators at the conference have set the fires on their way back home? He found out that of the 242 officers in attendance, fifty-five had left the conference alone and driven south along Highway 99. He decided to ask the FBI for help and rang Special Agent Chuck Galyan in Fresno. 'Fifty-five names of respected arson investigators? I thought Marv Casey was out in left field somewhere,' Galyan said. The case went cold.

Two years later, in 1989, there was another arson investigators conference in Pacific Grove, followed by another almost simultaneous outbreak of fires, this time along Highway 101, which hugs the coast from Los Angeles to San Francisco. Casey couldn't believe it. He worked out that only ten officers with southerly routes home had been at both the Fresno and Pacific Grove conferences. This time Chuck Galyan agreed to ask a fingerprint expert to check the relevant prints from the state database of public safety professionals. The veteran expert made his comparisons. But he failed to find a match.

Between October 1990 and March 1991 a rash of new fires broke out around Greater Los Angeles, in chain retail stores like Thrifty Drug Stores and Builders' Emporium. Glen Lucero, of the Los Angeles City Fire Department, said, 'The fires were occurring predominately during business hours. Most arson fires are set under the cover of darkness. This was highly unusual [and showed] a certain amount of bravado and confidence by the person setting the fires.'

In late March, the fires reached their apogee. Five stores were hit on a single day. The employees of one medium-sized craft store put out the blaze before it properly caught hold. Investigators found an incendiary device there, still in good condition, identical to the one Casey had found in Bakersfield four years previously. Six more of these devices were later recovered, a number of them in pillows, giving rise to the arsonist's nickname – the 'Pillow Pyro'.

The investigators knew they were after a clever, experienced and exceedingly dangerous man. He knew enough to start his store fires in the perfect place to encourage their rapid spread. People in these stores were in grave danger of meeting the same fate as those who had been trapped in Ole's Home Center in South Pasadena in 1984. That explosive fire had started in amongst polyurethane products, resulting in an inferno that burnt with a blue flame and an eerie hissing sound. Badly burnt bodies were blown out of the building by a flashover – when temperatures reach more than 500°C and all the combustible material in an enclosed space ignites explosively. Four people were killed, including a middle-aged woman and her two-year-old grandchild.

In April 1991, a 20-strong 'Pillow Pyro Task Force' was set up to liaise with police departments across California and track down their man. Three investigators visited Marvin Casey in Bakersfield, who eagerly showed them his photo of the fingerprint he had lifted in 1987. Because the print had already been cleared by an expert, the investigators had low expectations. But the Pillow Pyro might have committed a crime in the last four years, so they sent it to Ron George at the L.A. Sheriff's Department.

The Sheriff's Department database had a large collection of fingerprints, of criminals, of all police officers in the county, and of anybody who had ever applied for a police job. This time the examiner satisfied himself that he had a match – Captain John Orr, an arson investigator with twenty years' experience at the Glendale Fire Department behind him. Initially, the investigators couldn't believe that he was guilty, and clung to the idea that the fingerprint must have come from some sort of cross-contamination. On 17 April Ron George rang the Pillow Pyro Taskforce and told an agent, 'It's John Orr's. He shoulda known better. Tell that dummy not to handle the evidence.'

Orr's prints had been on the Sheriff's Department database since he was vetted for a job as a police officer with the LAPD in 1971. They had been happy with his prints, but not with a reference from his previous job which had described him as 'know-it-all, irresponsible and immature'. Further psychological tests confirmed his unsuitability for the role and they rejected him unceremoniously. Nevertheless John Orr's subsequent career in the Fire Service had been distinguished: he had personally instructed more than 1,200 firefighters, organised seminars on fire investigation and written a number of articles for the *American Fire Journal*. But how would John Orr have come into contact with evidence at a fire scene in Bakersfield, 100 miles from his base in Glendale?

There was only one unpalatable answer. The task force began surveilling Orr and talking quietly to some of his colleagues.

One of them had been suspicious for some time. He had noticed that Orr had an uncanny ability to arrive at a fire scene before anyone else, and rapidly home in on its origin. (As Niamh Nic Daéid explained earlier in this chapter, investigators work scenes in methodical phases, before approaching the business end.) But most of Orr's colleagues were incredulous. True, he could be smug when he talked about his investigations, but he was a damn good investigator, and one of their own.

Another conference was soon to be held at San Luis Obispo. The task force thought Orr might strike and wanted to catch him in the act. Agents watched him all weekend, every hour of the day, but he started no fires. It seemed he could feel their eyes on him.

In the end it was Orr's vanity that led to his downfall. He wrote a novel and sent it to a publisher with an astonishing cover letter. 'My novel, *Points of Origin*, is a fact-based work that follows the pattern of an actual arsonist who has been setting serial fires in California over the past eight years. He has not been identified or apprehended and probably will not be in the near future. As in the real case, the arsonist in my novel is a firefighter.' When investigators got their hands on it, they couldn't believe what they were reading.

The arson attacks mentioned in the manuscript matched many of the Pillow Pyro fires right down to the smallest detail, except for the names. The hero is a fire investigator on the hunt for the serial arsonist, Aaron. He compares the timings of all the fires to the working hours of the firemen in the force and realises that only Aaron could have done it.

On the morning of 4 December 1991 agents arrived at John Orr's home. Under the floor mat behind the driver's seat of his car, they found a pad of yellow lined notepaper. In a black canvas bag they found a pack of unfiltered Camel cigarettes, two books of matches, some rubber bands and a lighter.

The day after Orr's arrest, Mike Matassa of the task force made various calls to people he had worked with over the year. One was to Jim Allen, arson investigator and personal friend, who told Matassa, 'You ought to look at the Ole's fire. Y'know the one at Ole's Home Center in South Pasadena, October 1984? John's obsessed with that one. He was mad when they called it an accident.' When he got off the phone, Matassa had a flash of recall. Along with everyone else in the task force he had been reading a photocopy of *Points of Origin*. He remembered that in chapter 6 there was an account of a fire in 'Cal's Hardware Store', where five people had died, including a young boy. When Aaron doesn't get 'credit' for starting it he sets another fire in Styrofoam in a nearby hardware store, to expose the investigators' ignorance. The parallels were eerie.

On its own, *Points of Origin* would not have been enough to secure a conviction. But in conjunction with the other evidence – the fingerprint, and a tracking device that was installed behind the dashboard of his car – John Orr was found guilty of twenty-nine counts of arson and four counts of murder. He was sentenced to prison for life with no possibility of parole. He has never admitted to any of his crimes. But a fire investigator in *Points of Origin* makes a telling comment: 'The serial thing usually starts way after they have experimented with fire when they're young, and they just continue it if they aren't caught early. As they grow up, it takes on a sexual atmosphere. You know, they are too insecure to relate to people in a direct, person-to-person way and the fire becomes their friend, mentor and sometimes their lover. Actually a sexual thing.'

THREE

ENTOMOLOGY

'Augurs and understood relations have
By maggot-pies and choughs and rooks brought forth
The secret'st man of blood.'

Macbeth, III, iv

Our desire to understand how the dead meet their fate isn't a recent phenomenon. More than 750 years ago, in 1247, a handbook for coroners called *The Washing Away of Wrongs* was produced by a Chinese official named Song Ci. It contained the first recorded example of forensic entomology – the use of insect biology in the solution of a crime.

The victim had been stabbed to death by a roadside. The coroner examined the slashes on the man's body, then tested an assortment of blades on a cow carcass. He concluded that the murder weapon was a sickle. But knowing what caused the wounds was a long way from identifying whose hand had wielded the blade. So the coroner turned to possible motives. The victim's possessions were still intact, which ruled out robbery. According to his widow, he had no enemies. The best lead was the revelation that the victim hadn't been able to satisfy a man who had recently demanded the repayment of a debt.

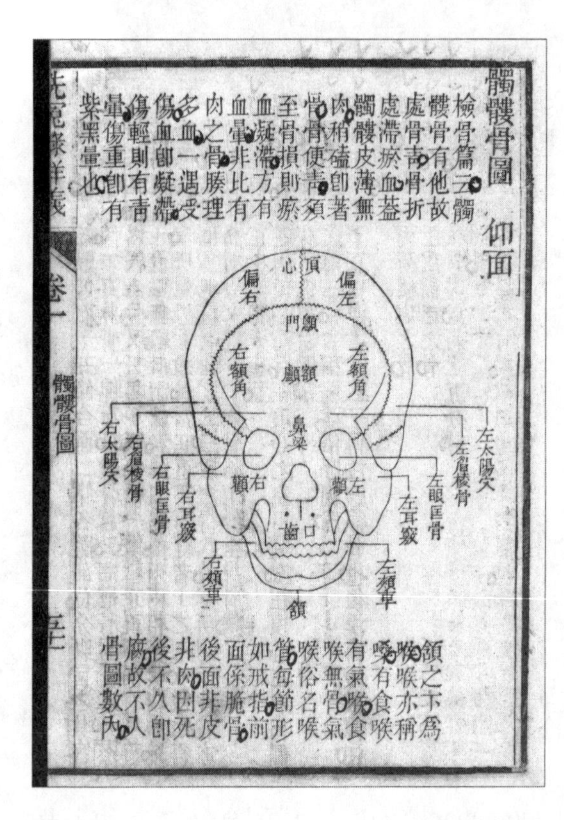

A page from a nineteenth-century version of The Washing Away of Wrongs, *a Chinese textbook on forensic medicine originally compiled by Sung Tzu in the thirteenth century*

The coroner accused the money lender, who denied the murder had anything to do with him. But the coroner was as tenacious as any TV detective. He ordered all seventy adults in the neighbourhood to stand in a line, their sickles at their feet. There were no visible traces of blood on any of the sickles. But within seconds a fly landed enthusiastically on the money lender's blade, attracted by minute traces of blood. A second fly followed, then another. When confronted again by the coroner, the money lender 'knocked his head on the floor' and gave a full confession. He'd tried to clean his blade, but his attempt to conceal his crime had been foiled by the insect informers humming quietly at his feet.

The Washing Away of Wrongs, the world's oldest extant book on forensic medicine, was updated and reprinted over 700 years, and carried to crime scenes by Chinese officials as recently as last century. When the first Portuguese traders reached China in the early 1500s, they were impressed by how reluctant the local courts were to condemn a person to death without an exhaustive investigation. The work of the modern forensic entomologist may be based on a wider and deeper range of knowledge, but it still epitomises the kind of careful scrutiny that impressed those Portuguese traders.

The usual role of forensic entomology in criminal investigations is to estimate time of death – a piece of information that's often crucial in establishing a suspect's alibi, and thus their guilt or innocence. The discipline is based on one grisly fact: a corpse makes a good lunch.

When forensic pathologists (see chapter 4) examine a corpse, they first try to estimate the time of death from phenomena such as rigor mortis, changes in body temperature and organ decomposition. After about forty-eight to seventy-two hours, these clocks have run down. But the time sequences provided by the insects arriving at the scene tick over long after that. Because different insects don't all turn up at the buffet together; there is a predictable order to their arrival. When the entomologists are called in they use their knowledge of this succession to estimate time of death. And so the insect kingdom helps the dead victims to provide unwitting but convincing evidence against their killers.

Most forensic entomologists don't start out with a passion for jurisprudence, but for insect life itself. And it takes years to develop the interpretive powers and expertise needed to bring the insect world to bear on criminal cases in a way that stands up in court. Yet the objectives of a passionate entomologist – to collect selectively and categorise scrupulously, to uncover the causes of curious behaviour, to find the evidence to prove theories – chime with those of a healthy justice system.

Jean Pierre Megnin was a key player in the development of modern forensic entomology. Like Song Ci, he wrote a surprisingly popular book, called *Les faune des cadavres* (*Fauna of Corpses*), published in 1893. Megnin recognised that hundreds of insect species are attracted to animal carcasses and, as a vet serving in the French army, he was perfectly placed to record the predictable waves in which they colonise the dead (which he detailed in an earlier book, *Faune des Tombeaux* (*Fauna of the Tombs*). He sketched many different species – particularly of mites and flies – at different stages of their development from maggot to adulthood, and published his illustrations for the general public.

The close observation and awareness of change over time that Megnin displayed set the tone for the emerging scientific discipline of forensic entomology. His scrupulousness gave the relationship between insects and the deceased an unprecedented legal status; Megnin was called upon as an expert witness in nineteen legal cases in France. And yet, entomology was still regarded as a largely anecdotal and haphazard branch of forensics. The principal problems were the range of variables entomologists have to consider – temperature, position of the body, variations in soil, climate and vegetation – and the lack of appropriate tools they had at their disposal in the nineteenth century. European and American scientists were nevertheless inspired by Megnin, and they worked throughout the twentieth century towards greater accuracy in identifying insect species and understanding the stages of their growth.

In 1986 Ken Smith, entomologist at London's Natural History Museum, wrote *A Manual of Forensic Entomology*, and dedicated it to Jean Pierre Megnin. His book was a game-changer. Smith pulled together all the available information on carrion-loving insects, particularly flies, and showed more accurately than ever before how to use them to age corpses. The manual was practical, something to be taken to the scene of an investigation. It described the waves of species as they appeared on corpses that were buried, exposed or submerged in water. Smith was an outstanding taxonomist, too, producing identification guides which are still used today. Reading the *Manual* in conjunction with the guides, it became possible to determine where flies had first found a corpse, even if it had subsequently been moved.

Ken Smith's successor at the Natural History Museum is Martin Hall. He's a tall man who strides through the museum's galleries giving a cheerful and enthusiastic running commentary. His passion for the 30 million insect specimens he curates is obvious and infectious.

He juggles his day job at the museum with the role of forensic entomologist. At any time his mobile phone might ping with a police request to drop everything and rush to a crime scene. 'Collecting insects off a dead body is not a pleasant experience,' he says, 'but it is amazing how your professional interest takes over.'

Martin's fascination with his subjects was sparked when he was a boy, growing up on the east African island of Zanzibar. There, he realised that the mosquito net that hung above his bed could keep bugs inside his world even better than bar them from it. As he dropped off to sleep every night, stick insects, praying mantises, even the occasional bat crawled, buzzed or flew their way through his semi-consciousness.

He went to England to study before returning to Africa and spending seven years researching tsetse fly behaviour. One day he saw the massive corpse of an adult elephant on the savannah, its flesh creeping with countless maggots. A week later he came back to find nothing but a giant skeleton, stripped bare. Another week, and blowflies swarmed the site like low rain clouds. 'It was just extraordinary to watch. Although there were other scavengers, like hyenas and vultures, the maggots were probably responsible for 40 to 50 per cent of the biomass loss.' An elephant had become a million flies, and a budding entomologist was hooked for life.

Now, that zest for his work is communicated to everyone Martin meets. When I visited him at the museum, he whisked me behind the scenes and up dozens of stone steps to the very top of one of its high gothic towers with a panoramic view of London. But I wasn't there for the view. I was there to see the setting for some of the experiments Martin and his team conduct to extend the range of their knowledge. It's a world where familiar objects have quite different meanings. Carry-on suitcases are home to pigs' heads, in order to explore which flies will manage to lay their eggs through the gaps in the zips. Dog cages hold rotting piglets. Tupperware sandwich boxes are filled with preserved maggots. It's all more than a little unsettling. Small wonder I later refused his offer of a sandwich …

Among the collection of insects in the museum are some that have historic significance. Martin showed me one specimen bottle, then said in hushed tones, 'These are iconic maggots. They're from the Buck Ruxton case.'

The Buck Ruxton case is notorious in British criminal history. It was a landmark case for forensic science in several respects, but forensic entomologists like Martin Hall know it as the first case in the UK in which insects were successfully used to help solve a crime. The case was a sensation, filling column after column of newsprint in the autumn of 1935.

Buck Ruxton was a doctor of Parsee and French extraction, who had qualified in Bombay and settled in northern England. He lived with a Scottish woman called Isabella, who people knew as 'Mrs Ruxton', and their three young children. The doctor was the first non-white medical practitioner in Lancaster, and was very popular, especially with his poorer patients.

One Sunday morning, Dr Ruxton opened his front door to a scrawny 9-year-old boy. His mother stood behind him expectantly, her arms shielding him from the autumn cold. 'I'm sorry,' said the doctor. 'I can't perform the operation today, my wife has gone away to Scotland. There is just myself and my little maid here, and we are busy taking the carpets up ready for the decorators in the morning. Look at my hands, how dirty they are.' The pair turned and walked away disconsolately, the mother wondering to herself why the single hand the doctor had held out looked so clean.

The Ruxton family had a 19-year-old maid, Mary Rogerson. A few days after the incident on the doctor's doorstep, her family reported her missing. The police visited Dr Ruxton, who claimed his wife had gone off to Blackpool with her maid and that he suspected Isabella of having a lover. That fitted with the last known sighting of Isabella, who had been seen driving away from Blackpool at 11.30 p.m. after an evening with friends. Her love of having a good time provoked blazing rows between the Ruxtons. Dr Ruxton constantly accused his wife of infidelity and Mary often witnessed his jealous rage.

After the police visited him a second time, Ruxton claimed that Isabella and Mary had gone to Edinburgh. But he couldn't stop the tongues wagging in Lancaster. Although Ruxton was a respected member of his community, the gossip spread that his rows with his wife had grown more harsh and bitter over the summer and that something more sinister might be behind the disappearances.

Then, on 29 September, a woman was walking across a bridge over a ravine near Moffat on the road from Carlisle to Edinburgh when, horrified, she realised she was looking at a human arm sticking up from the bank of the stream below. When the police arrived on the scene they found thirty bloody packages containing body parts wrapped in newspaper. Over the next few days, other body parts were found in the area by police and members of the public. Seventy parts were eventually recovered, from two different corpses. They had almost certainly been butchered to prevent identification – the fingertips had been cut off – and the job had been done by someone who knew about human anatomy.

Some maggots were found feeding on the decomposing parts, and were sent off to the University of Edinburgh. There the entomologists identified them as a particular kind of blowfly. They narrowed down the time since the body parts had been dumped as between ten and twelve days. And so police linked the parts to the disappearance of Isabella and Mary.

It was a telling start, but the evidence against Buck Ruxton stretched far beyond maggots. An anatomist and a forensic pathologist from Glasgow and Edinburgh universities painstakingly reconstructed the bodies of the victims. They superimposed photographs of the living Isabella on to photographs of one of the skulls, which matched. Some of the body parts had been wrapped

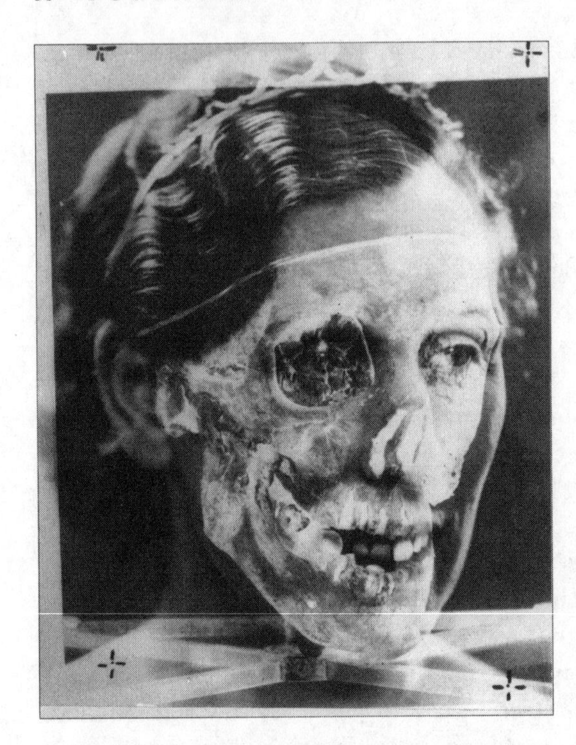

This image, which superimposes a photograph of Isabella Ruxton's face onto the skull found in the stream, helped seal Buck Ruxton's conviction

in a special pullout section of the *Sunday Graphic* newspaper, distributed only in the Lancaster/Morecambe area, on 15 September. Some were wrapped in clothes belonging to the Ruxton children.

Ruxton had clearly been less composed than he had hoped. In his hurry to get away from the ravine and back to Lancaster, he had knocked a man off his bicycle in his car. The rider had scribbled down the number plate. It led straight to the car Buck Ruxton owned. The date of the reckless incident dovetailed perfectly with the evidence of the maggots and *Sunday Graphic*.

The final piece of the jigsaw came from local knowledge. The stream in the ravine had last flooded on 19 September. The bodies must have been there by that date because some parts, such as the ghoulishly raised arm, had been deposited on the bank at a level where the overflowing stream had reached.

Buck Ruxton was arrested and found guilty of murder. Nine months after committing his crimes, he was hanged at

Strangeways Prison in Manchester. We will never know the exact circumstances of what became known as the 'Jigsaw Murders'. But, based on the autopsy evidence, it is most likely that Ruxton strangled his wife with his bare hands. The maid died from having her throat cut, probably to silence her after she discovered his crime.

The insect evidence was just one tile in a forensic mosaic that spelled out the murderer's guilt. But the success of the combination of methods used in the Ruxton case led to increased public and professional trust in the capabilities of forensic science, including the discipline of forensic entomology. People could see that even if Buck Ruxton had wrapped the diced parts of his victims in white paper bags rather than sections of the local newspaper, even if his car hadn't hit a bicycle, even if the stream hadn't burst its banks, the maggots would have pointed to him. And so fresh minds were drawn to the discipline.

Martin Hall has devoted much of his life to blowflies – the family of insects most commonly associated with corpses. There are over a thousand known species in the world. Martin considers blowflies 'the gold standard indicators' of the forensic world, for several reasons. Because of their acute sense of smell, which can pinpoint the tiniest drop of blood or the merest whiff of decomposition at over a hundred metres, they colonise dead bodies more quickly than other insect families. Because so much is known and documented about the stages of their growth, they usually provide the best information for gauging minimum time elapsed since death. And because there are so many regional varieties, they can be used to plot the whereabouts of a murder scene even when the body is found elsewhere.

Unlike other insect families, which use their sense of smell only until they get close enough to food to switch to eyesight, the blowfly uses its sense of smell right up until it lands on what it has smelt. That makes it very difficult to dispose of a body in such a way that blowflies won't find it. If, for example, a body is hidden underneath the floorboards, the odours of decomposition will gradually percolate through the airbricks, and the flies will crawl through them to find the body.

Even if a body can be hermetically sealed off, clues to its location may still be obvious. American police in Indiana searching for a missing person some years ago noticed a cloud of frustrated flies hovering above a covered well. The missing person had been murdered, and the killer had thrown their remains down the well. He had sealed the well up enough to prevent insects getting in, but not the slight odour of decomposition getting out. The flies acted like a swaying gravestone, drawn to a smell beneath that was far beyond the capabilities of the human nose.

Not long after Barack Obama took office in 2009, a blowfly buzzed around his head during a live interview on CNBC. The fly finally landed on the back of his left hand, and he promptly killed it with a slap from his right. 'That was pretty impressive, wasn't it?' he said. 'I got the sucker.' In 2013 another fly landed on the president, this time right between his eyes. It made a good photograph. But when Martin Hall sees something like that, his mind is already racing ahead to what those flies would have done if the president hadn't been in a position to swat them. 'They would have gone on to explore his body. If they were females with a pile of eggs ready to be laid, they would look for a suitable site, usually the head orifices, in the nose and the eyes and the mouth. And they'd lay their eggs.'

And then the banquet would begin. In 1767 Carl Linnaeus, the father of modern taxonomy, observed that 'three flies consume the corpse of a horse as quickly as a lion does'. This startling observation came to Linnaeus because of the pioneering work of Francesco Redi. In 1668 the Italian had proved with a series of experiments that maggots came from fly eggs. Before Redi, the presence of maggots in corpses was presumed to be the result of spontaneous generation.

Once a female blowfly has laid her eggs, a biological clock begins to tick. In the height of summer a typical UK blowfly egg will take fifteen days to become a fly. After one day the egg hatches into a maggot, which shreds and rakes in decaying flesh with the two hooks on its mouth. Because its eating and breathing organs are at opposite ends of its body, it can eat and breathe simultaneously twenty-four hours a day. Over the next four days it eats voraciously and grows to ten times its original size, from 2 mm to 2 cm.

The plump maggot then wriggles away from the body, towards a dark place where it's less likely to be eaten by a scavenging bird or fox. If its fleshy nursery happens to be outdoors, it will burrow 15 cm down into the soil. If indoors, the underside of a wardrobe or a place between the floorboards does the job. Secure in the darkness, the maggot becomes a pupa, its third and final outer layer of skin hardening into a case. Ten days later, an adult fly breaks out of the case and, if it's outside, tunnels up to the surface of the soil. This push for freedom is no mean feat. The fly fills a sac on its head with blood, and pulsates its balloon-shaped battering ram inwards and outwards to dislodge the soil. Once it reaches the air, the fly shakes out its crumpled wings, and begins to mate almost immediately. At two days old a female lays her eggs, sometimes on the same corpse that reared her – but because maggots can devour 60 per cent of a human body in under a week, there probably won't be much left.

In the woodlands, bedrooms, alleyways and beaches that the police call him to, Martin Hall encounters the strange music of hordes of flies. He sees sights and smells of undeniable variety. 'Sometimes you hear the description of "the sweet smell of decay", and it can be sweet at times, but it can also be quite overpowering. I've worked on cases where the upper half of the torso is completely skeletonised because it was protruding from a sleeping bag and the lower half doesn't look that long dead. As we approach the scene it's not that bad but as soon as the bag's opened, the waft hits you. The smell is not just of the body, but of the maggots feeding on the body. They produce a lot of ammonia and that can be overpowering, too.'

◆ ◆ ◆

Sometimes crime scene investigators collect insect specimens from corpses, and send them to entomologists to inspect. Martin Hall prefers to visit the scene himself. That way he can ensure that the specimens and information gathered are admissible in court, and he has the opportunity to search in places that others might miss or simply not consider. He looks for maggots throughout the corpse and for pupae under the soil. He wants to find the oldest specimens because they reveal when the flies first found the corpse, and so indicate the minimum time elapsed since death. Martin kills some of those maggots in boiling water and stores them in ethanol. He keeps others alive. Maggots grow faster the warmer it is, so Martin puts down a thermometer box which logs the temperature on the hour every hour for the next ten days. He also gets readings for the past couple of weeks from the weather station closest to the scene, so he knows roughly how hot the maggots were as they grew.

Back at the lab Martin makes the crucial identification of the preserved maggots. 'Even closely related species develop at different rates, so if you get it wrong you can be giving the police wrong information.' He incubates the live maggots until they become flies as confirmation of his identification. He examines the anatomy of the preserved maggots carefully to assess their stage of growth. Combining his assessment with the temperature data, he plots a graph leading back in time to the point when the mother blowfly laid her eggs. This is usually a key piece of information, and an entomologist's most valuable offering to the forensic puzzle.

But what if the corpse has been there for longer than seven days, which is roughly how long it takes a maggot to become a pupa? Can entomologists look further back in time than a week? As entomologists start to push the boundaries of what insects can tell us, they are discovering how to read the biological clock embedded in pupae, too.

It takes ten days for a pupa to transform into an adult fly. It's this process of metamorphosis which is at the heart of what makes insects mysterious, and has been provoking wonder in poets and entomologists alike for centuries. It hasn't been possible to witness pupae as they change over time, because their case is opaque. With the help of X-rays and miniature CT scanners, though, Martin and his team at the Natural History Museum are changing that. Having helped to plot reliably the growth rates of many species of blowfly maggot, he is now concentrating on the art of aging pupae: 'At thirty hours I took an X-ray image of a specimen and it was just the larval [maggoty] tissues. I took an image just three hours later, when I came back after a cup of tea, and that specimen had completely transformed. Instead of this undifferentiated larval tissue you could now see a clear head, thorax, abdomen and the developing legs and wings.'

It's tempting to think that, armed with such astonishing insights, forensic entomologists are beginning to deal in the currency of pinpoint certainty. But jurors and entomology students should not to be seduced. In 1994 a cartoon accompanied a BBC programme called *The Witness was a Fly*, in which a magnifying glass that could have belonged to Sherlock Holmes hovers above a maggot. The maggot holds up a placard saying 'Murdered, 3 p.m. Friday'. The cartoon is striking but it's also misleading. Maggots cannot tell us when a murder took place. They can indicate when flies laid eggs on a corpse, and that reveals the point by which the person was definitely dead. In the warm summer months it would be possible to narrow that window to, say, Friday, and possibly, as deductions become increasingly refined, to Friday afternoon. But to expect an entomologist to give a definitive time of death to the hour would be like asking a weather forecaster in November to promise a white Christmas. The range of variables thwarts that level of accuracy.

One of those variables is based on maggots' gregariousness. They like to feed in 'maggot masses' – a sort of seething mosh pit. They lay down an alkaline residue as they move about, which breaks tissue down into ammoniac goo. Their digestive activities are so intense that they heat carcasses up, sometimes to as high as 50°C. This suits the blowfly family because warm conditions accelerate their growth, but it can be a headache for entomologists trying to map their activity. However, it's only in the latter stages of development that maggots generate significant heat. So the earlier the entomologist gets to the maggots, the lesser the effect of the maggot mass.

If the oldest maggots can't be found, because they are already flies, then the entomologist can draw on Jean Paul Megnin's nineteenth-century insights about the predictable waves of insect colonisation. As a corpse begins to dry, different fly families, such as cheese flies, flesh flies and coffin flies, make it their home. When a corpse becomes too dry for a maggot's mouth hooks to rake, beetles arrive armed with chewing mouth parts. They eat the dry flesh, skin and ligaments. Finally, moth larvae and mites go to work on the hair, leaving only a skeleton to signify the life that once was. All of these species work to their own timetables, which the entomologist can harness to help estimate the length of time since death.

In 1850 a plasterer discovered a mummified child behind a mantelpiece in Paris. The young couple living there were initially suspected of murder but, when Dr Bergeret d'Arbois looked at the insect evidence, he contended that the body had been 'exploited' by flesh flies (which differ from most other flies in that they are ovoviviparous – that is, they deposit hatched maggots, rather than eggs, into decaying matter or open wounds) in 1848, and mites had laid eggs on the dried corpse in 1849. Suspicion then fell on the former occupants of the house, who were arrested and subsequently convicted.

In some cases, the puzzle faced by the investigators is nothing to do with the time of death. In a recent case in Merseyside, police were searching a suspect's house when they came across a large collection of empty pupal cases. They speculated that these could have been the result of a dead pigeon in the loft, but it seemed odd that there were the remains of so many pupae. But there was no way of interrogating the dark brown cases to discover when they had cracked open to free the newborn flies. Then someone had the bright idea of sending the cases off for a toxicology examination. The results were startling. The cases contained traces of heroin metabolites. There's no history of pigeons ingesting heroin, so further tests were ordered. Martin explains: 'Maggots feed in a soup of DNA, and they have spines on their bodies which tissue gets lodged in. The pupal case is the old skin of the maggot and may still have human tissues on it.' When the pupal cases were examined further they revealed traces of human DNA which matched that of a known drug user who had been reported missing. On the basis of this and other evidence, the owner of the house was convicted of murder and jailed for life. He had disposed of his victim, but could not silence the insect witnesses.

More conventionally in forensic entomology, the time of death can sometimes play a decisive role in a court case. One day in the park, a 10-year-old British girl, Samantha, met a man of about thirty. He gave her sweets and befriended her. When she got home Samantha told her mother what had happened. Her mother didn't seem too concerned about her daughter's new acquaintance. A while later the girl met the man again and this time he invited her back to his house. Nothing terrible happened. The pair met like this regularly. They would go for walks, or watch TV, sometimes with a few of the man's male and female friends there, too. Eventually the girl invited the man back to her mother's house. Before long, the mother and the man began having a relationship. The man spent some weeks with the mother, before beginning to abuse Samantha sexually. The household filled with bitter resentment. Vicious rows flared up between the three of them. And then Samantha went missing.

The police conducted a search and finally uncovered the girl's body in a mound of rubble and broken bricks in the grounds of a hospital. A ferocious blow from a heavy blunt object had caved in the left side of her skull. The distinguished forensic entomologist

Zakaria Erzinçlioğlu was summoned to the scene to examine the body. He discovered some freshly laid eggs and minuscule blowfly maggots. These offered evidence that the girl had died very soon after she was last seen with the man. In court the man pleaded not guilty. But halfway through the proceedings, as the maggot evidence was being presented, the man broke down and confessed. He had silenced the girl in the middle of an argument, after she had threatened to tell her mother about what he was doing to her.

Zak Erzinçlioğlu has helped to solve 200 homicides in his 30-year forensic career, and has written about many more, but his quirky memoirs, *Maggots, Murder and Men* (2000), cover much more ground than the title suggests. For example, he records a peculiar incident in Finland. One morning, a government official walked into his office to find some large maggots underneath the edge of the carpet. He summoned his cleaning lady and asked her when she had last cleaned his office. When she said 'last night', he fumed at her and accused her of lying. He simply couldn't believe that such big 'bugs' could have appeared overnight. He fired her on the spot.

But the official, a typical bureaucrat, preserved some of the maggots and eventually took the opportunity to show them to a professor at Helsinki University. He identified them as blowfly maggots in the wandering stage. They had finished feeding, probably on a rat elsewhere in the building, and wriggled off to find somewhere to pupate. They could indeed have arrived in the office overnight. The mortified official contacted his old employee in a state of remorse, offering to restore her to her former job.

In the end, it all comes down to placing science at the service of justice. It's about taking facts which have been hard won in the abstract world of the laboratory and using them in the uncompromisingly real world of the crime scene. 'In an academic environment you don't encounter these sort of things,' explains Martin. 'But there is a great satisfaction to be got from applying my knowledge of insects to help in a fairly short timespan. Many scientists, not just entomologists, beaver away for years and years and years and don't necessarily see an outcome from what they do; whereas I can generally see within a few months that something I've done has actually been of help.'

Martin also recalls one case in Yorkshire, in which an elderly man had sold all his fine antique furniture for next to nothing to a stranger who had talked his way into his house. The confidence trickster told the old man that the furniture was infested with woodworm, showed him maggots on the floor to prove it, and left with his booty. Distraught, the old man called on his neighbour, who spotted some of the maggots still on the floor. He bottled them up and gave them to the police, who in turn passed them on to Martin. The maggots were promptly identified as crane fly larvae – better known as Daddy Longlegs – which like to feed on grass roots and are completely uninterested in wood.

Martin said, 'Fortunately, the guy who stole the furniture was found and it was returned to the old man. Even this brusque Yorkshire policeman was quite emotional when he was telling me how excited the old guy was when he got his stuff back. Again it was just through knowledge of insects.' So happy endings rooted in knowledge are out there; but entomological evidence can be a slippery thing, especially in the adversarial world of the courtroom.

On Friday, 1 February 2002, Brenda Van Dam went to a bar with two friends in San Diego, California, leaving her husband to babysit their three children. Brenda arrived back home at 2 a.m. She only noticed that her 7-year-old daughter, Danielle, was missing from her bedroom when she went to wake her up the next morning. Panic seized her. The last time she had seen Danielle was the previous evening, when the little girl had been writing her diary, while her father and brothers played video games.

The police interviewed neighbours and discovered that David Westerfield, an engineer who lived two doors down from the Van Dams, had driven his motorhome away for the weekend.

David Westerfield stands during a plea hearing in a San Diego Superior Courtroom in February, 2002. He pled not guilty to a charge of murdering Danielle van Dam, his seven-year-old neighbour

Everybody else in the neighbourhood had been at home. It emerged that Danielle and her mother had knocked on Westerfield's door a few days earlier, selling Girl Scout cookies. He was placed under 24-hour police surveillance on 4 February. Police investigated his motorhome and found child pornography, some of Danielle's hair, fingerprints and a smudge of her blood. After a search involving hundreds of volunteers, Danielle's naked body was found on 27 February in a dry, brushy area by the side of a road. Her shrivelled, leathery skin had almost entirely mummified.

Insect evidence became a key focus of the trial. An unprecedented *four* entomologists were called to testify. Very few maggots had been found inside Danielle's body. The entomologists called by the defence said that flies must have laid their eggs in mid-February. The defence lawyer argued that, as this was a week after Westerfield had been placed under police surveillance, he couldn't have dumped her body at the roadside. The prosecution lawyer accused the defence entomologist of using incorrect weather data. He sneeringly asked him how much he was being paid, implying that he was a 'hired gun', which led to such a blazing row that the session had to be adjourned early.

Entomologists called by the prosecution dated the blowfly infestation to 9–14 February, still some days after Westerfield was put under surveillance on 4 February. But, they argued, other factors might explain the late arrival and small number of insects found on the corpse. The extremely dry weather – the driest in more than a century – had sucked the moisture out of Danielle's body, making it less attractive to maggots. A blanket may have been covering the body which could then have been carried away by dogs. Perhaps ants had taken away the eggs and maggots that had arrived earliest to the corpse. These ideas were refuted by an entomologist for the defence.

David Westerfield was found guilty of kidnapping and murder. And put on Death Row. The average wait between sentencing and execution is sixteen years in California. He protests his innocence to this day. In 2013, he made an official appeal for a new trial, which has yet to be heard by the Supreme Court.

The conflicting conclusions of the four experts involved in the Westerfield case did damage to the reputation of forensic entomology. There is no evidence that any of them were 'hired guns'. Rather, they had a particularly difficult set of circumstances and variables to deal with: a small number of maggot specimens; conflicting reports about the extreme weather; intense media scrutiny. Only one of the entomologists was actually given the chance to examine the body *in situ*. Science usually works best when it is approached collaboratively. If those scientists could have compared their findings in an unpressured environment – as often happens now in the UK system – the range of their estimations would likely have been reduced.

Since the Buck Ruxton 'Jigsaw Murders' in 1935, the UK public's appreciation of forensic entomology has grown steadily. And now, thanks in part to the international success of crime television shows such as *CSI*, whose main character, Gil Grissom, frequently uses insects to solve crimes, forensic entomology is more widely recognised than ever. Real life entomologists continue to dream up astonishing ways of using their science to extract forensic evidence. In a recent case in the US, the movements of a suspect were identified by the insects splattered on the windscreen of his vehicle.

But such breakthrough techniques are not the norm. Most of the forensic entomologists' work is based on a detailed assimilation of a huge volume of information and the ability to differentiate between insects that most of us would dismiss as identical. Entomologists turning to forensics enter an emotionally and intellectually complicated place. Their knowledge and methods are stretched to the limit as they try to read minute biological clocks under their microscopes. Getting informants to hand over their secrets is a tricky business, no matter what their species.

FOUR

PATHOLOGY

'To begin depriving death of its greatest advantage over us, let us adopt a way clean contrary to that common one; let us deprive death of its strangeness; let us frequent it, let us get used to it; let us have nothing more often in mind than death.'

Michel de Montaigne, *Essais* (1580)

The poet John Donne reminds us that 'Any man's death diminishes me, because I am involved in mankind'. It's a line that carries moral weight but, in spite of that, it is undeniable that we are more affected by sudden violent death when it has some connection to our own lives, however tangential. So it is for me with the case of Rachel McLean, who was an undergraduate at the same small Oxford women's college that I attended. Although I never knew her, I can't escape a sense of distant kinship with her and her fate.

Rachel McLean was an undergraduate at St Hilda's College when she became John Tanner's girlfriend at the age of nineteen. Tanner proposed to her on 13 April 1991, ten months into their relationship. A momentous occasion like that would surely be something any girl would be eager to talk about with everyone who is close to her. But over the next few days none of her friends at St Hilda's or in the wider university saw Rachel. She was both a studious and a friendly, open person; nobody could quite believe she'd have gone off somewhere without telling anyone. Tanner phoned the house where she lived, saying he wanted to speak to her but her housemate said she didn't know where Rachel was.

After five days of growing concern, the college authorities reported Rachel missing to the police. When they contacted Tanner in Nottingham, where he was studying at the university, he explained that he'd last seen her on the evening of 14 April 1991, when she had waved goodbye to him from the platform at Oxford station as his train left for Nottingham. A long-haired young man they'd met in the station buffet had offered her a lift back to her digs in Argyle Street.

Tanner co-operated with the police, helped in the searches and took part in a television reconstruction of his departure from Oxford station to try and jog the memory of anybody who might have seen Rachel. He's believed to be the first murderer ever to take part in such a televised reconstruction. In an emotional press conference he told friends and reporters that he and Rachel were in love, and planning to get married.

But the police suspected that Tanner was hiding something, so they had briefed reporters to ask him key questions, such as 'Did you kill Rachel?' The way he answered, smirking and with a lack of emotion, made the police feel sure he knew more about her disappearance than he was willing to admit.

They searched the house in Argyle Street that Rachel shared with friends. Everything appeared to be in order; the floorboards hadn't been tampered with and nothing else looked suspicious. The detectives were desperate to find evidence that they could use to arrest Tanner, or at least to put pressure on him. Frogmen searched the River Cherwell and other officers scoured nearby scrubland.

They contacted the local council to check whether the house on Argyle Street had ever had a cellar. The answer came back that, although there was no basement, some of the houses on the street had been underpinned, which meant that there were spaces underneath the floors.

Armed with this information, the police searched the house again on 2 May. This time they found Rachel's partially mummified body under the stairs. Tanner had squeezed her through a 20 cm gap at the bottom of the stair cupboard, and pushed her underneath the floorboards. Although it had been eighteen days since her death, she had barely decomposed; warm, dry air coming through the airbricks had dried her skin.

Finding the body is the end of the first phase of a murder inquiry. But it's only the beginning of the forensic pathologist's crucial contribution to the building of a case against the defendant. In the Rachel McLean case, that task fell to Iain West, head of the Department of Forensic Medicine at Guy's Hospital. During the autopsy, he found a 1 cm bruise to the left of Rachel's larynx and four 1 cm bruises to the right. He took photographs of these, and of petechiae – tiny haemorrhages in her face and eyes. His internal examination revealed fractures of the laryngeal cartilages in the throat. All of these traumas were indicators of death by strangulation. There was also a tuft of hair missing from her scalp, which West believed Rachel had torn off in a desperate attempt to relieve the pressure on her throat.

When the police confronted John Tanner with the damning evidence from Iain West, he broke down and confessed to killing Rachel. At his trial he said, 'I flew at her in a rage and proceeded to put my hands around her neck. I think I must have lost control, because I have only a vague recollection of the time that elapsed afterwards.' He claimed he had killed Rachel after she had admitted to being unfaithful. Then he spent the night lying next to her lifeless body. In the morning he searched for a suitable place to hide her, dragged her through the gap in the cupboard, and caught a train back to Nottingham. Tanner was given a life sentence. He returned to his native New Zealand upon his release in 2003.

Forensic pathology resembles a jigsaw puzzle. The pathologist has to catalogue any unusual elements found on or inside a corpse and, from those fragments of information, try to reconstruct the past. Throughout human history, people have wanted to understand why those they care about have died. The very word 'autopsy' derives from the Ancient Greek for 'seeing for oneself'. An autopsy is a medical attempt to satisfy that profound curiosity.

The first known forensic autopsy took place in 44 BC, when Julius Caesar's doctor reported that, of the emperor's twenty-three stab wounds, only the one between his first and second ribs was fatal. A couple of centuries later, the Greek physician Galen produced hugely influential reports based mainly on dissecting monkeys and pigs. Despite his raw material, his theories on human anatomy remained uncontested until Andreas Vesalius started comparing normal and abnormal anatomy in the sixteenth century, paving the way for modern pathology, the science of disease.

When Vesalius published his landmark book on anatomy, *De Humani Corporis Fabrica* (*On the Fabric of the Human Body*), in 1543 he dedicated it to the Holy Roman Emperor, Charles V, whose reign also saw another landmark in forensic medicine. For the first time in the history of the Holy Roman Empire, rules of criminal procedure were enacted. They regulated which crimes should be regarded as serious, allowed for the burning of witches and, for the first time, gave the courts the power to order investigations and inquisitions into serious crime. Known collectively as the Carolina Code, crucially for forensic medicine, they required judges to consult surgeons in cases of suspected murder.

The Code was adopted across much of continental Europe and medical authors grew keen to display their expertise in the courtroom. Those authors included the French barber surgeon Ambroise Paré, sometimes called 'the father of forensic pathology'. He wrote about the effects of violent death on internal organs, explained the indications of death by lightning, drowning, smothering, poison, apoplexy and infanticide, and showed how to distinguish between wounds made on a living and a dead body.

◆ ◆ ◆

As our understanding of the workings of the human body grew, so did the discipline. In the nineteenth century Alfred Swaine Taylor wrote extensively on forensic pathology and modernised the discipline in Britain and abroad. His most important textbook, *A Manual of Medical Jurisprudence* (1831), went through ten editions in his lifetime. By the mid-1850s Taylor had been consulted on more than 500 forensic cases, but his experience provides an early demonstration that forensic scientists are as fallible as the rest of us.

In 1859 Dr Thomas Smethurst was tried at the Old Bailey, charged with poisoning his mistress, Isabella Bankes. At the trial Swaine Taylor testified that there was arsenic in a bottle owned by Smethurst, and that this was proof of his guilt. Smethurst was found guilty, and sentenced to death. It later became clear that Swaine Taylor had carried out the test inadequately and that the bottle was in all probability arsenic-free. Isabella Bankes had been ill for a long time before, and was most likely to have died of natural causes. Smethurst was pardoned, though he did have to serve a 1-year prison sentence for bigamy. The *Lancet* and *The Times* heavily criticised Swaine Taylor and the verdict of murder, and forensic pathology became known as 'The Beastly Science'. The case cast a heavy shadow over forensic pathology for many years.

Given the theatrical nature of the adversarial court system, where the two sides of a case are represented by advocates before an impartial judge, what the discipline needed to restore its reputation was someone whose ability came with a liberal sprinkling of charisma. That touch of glamour arrived in the person of Bernard Spilsbury. A handsome, convincing orator, Spilsbury was never seen in public without a top hat, tails, buttonhole and spats. His skill was evident. His ambidextrous hands could work on a dead body with speed and precision. He also presented his findings in clear, everyday language.

Juries and the public loved Spilsbury. The press portrayed him as a steady rock on which the law could break open the lies of immoral murderers. On his death in 1947 the *Lancet* said he had 'stood alone and unchallenged as our greatest medico-legal expert'. Spilsbury had appeared for the Crown in over 200 murder cases.

Dr Hawley Crippen and his lover, Ethel le Neve, in the dock at the Old Bailey. Crippen would be convicted of murder and sentenced to death, while le Neve walked free

The first time he came to the attention of the public was in 1910, as an expert witness at the sensational trial of Doctor Hawley Harvey Crippen. An American homeopath and patent medicine salesman, Crippen had been living in Camden Town with his wife, Cora, a music hall singer with the stage name 'Belle Elmore'. The marriage had been in difficulties; then Cora's friends suddenly stopped seeing her around. Doctor Crippen told them variously that she had died and that she had gone to America to further her career. They became suspicious and went to the police, who questioned Crippen and searched his house. They found nothing. But the investigation panicked Crippen, who fled with his teenage mistress, Ethel Le Neve, aboard the *SS Montrose* bound for Canada. Le Neve dressed as a boy and posed improbably as his son.

Their flight rekindled the suspicions of the police, who went back to search the house a second time, and again found nothing.

But they remained suspicious and mounted a third search in which they dug up the brick floor of the cellar. This time, they found what were believed to be the remains of a human torso, wrapped in a man's pyjama top.

Meanwhile, the captain of the *Montrose* had noticed two odd passengers on board, and, just before his vessel went out of range, he sent a wireless telegram to the British authorities: 'Have strong suspicions that Crippen London cellar murderer and accomplice are among saloon passengers. Mustache taken off growing beard. Accomplice dressed as boy. Manner and build undoubtedly a girl.' Chief Inspector Dew of Scotland Yard boarded a faster ship, landed in Canada before the pair, and dramatically arrested them when they arrived – the first arrest made with the help of wireless communication.

The police called in a surgeon from St Mary's Hospital, London, to examine the body and he put the young Spilsbury on the case. Spilsbury took a close look at Cora Crippen's medical records and noticed that she had had an operation on her abdomen. The autopsy did not reveal the sex of the body, but Spilsbury did find traces of toxic compounds.

At Crippen's trial, Spilsbury presented a piece of skin bearing a curved scar, preserved in formaldehyde, from the torso believed to belong to Cora Crippen. He passed it around the jury in a glass dish. Spilsbury set up a microscope in the next room for members of the jury to examine slides of the tissue. Although the defence pathologist argued that, because it had hair follicles growing from it, it must be folded skin and not scar tissue, the jury believed Spilsbury. Crippen was found guilty of drugging and murdering his wife. He was hanged at Pentonville Prison in London and at his request buried with a photograph of Le Neve. She was charged, and then acquitted, of being an accessory after the fact.

The Crippen slides still exist, in the Royal London Hospital, and in 2002 Professor Bernard Knight examined them. He could not see definite indications of scar tissue. Recent DNA testing of the fragments seem to show that the DNA does not match that of some of Cora Crippen's descendants, and that the remains are

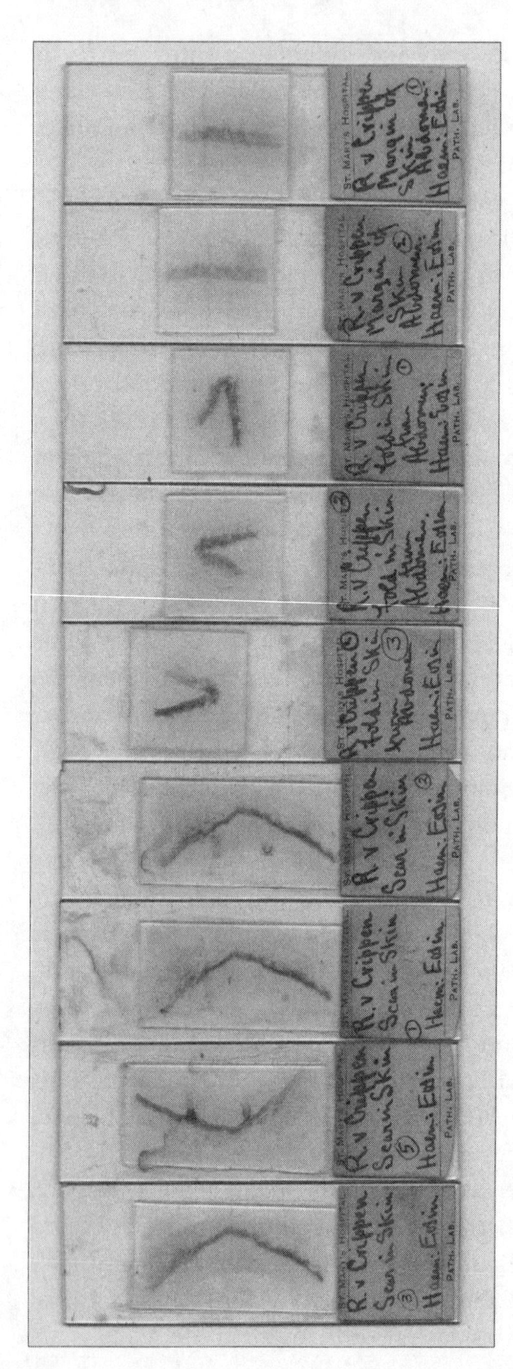

A series of slides produced by Spilsbury showing sections of the scar found on the torso buried in the cellar: according to him, this proved it belonged to Cora Crippen, though others disagreed

male. Ironically, it seems that the case which projected Spilsbury into the public consciousness as the standard bearer for forensic pathology may have been one in which he was comprehensively mistaken.

Five years after Crippen was hanged, Spilsbury was involved in another extraordinary case that neither DNA testing nor any other modern forensic technique could have helped him solve. On Sunday, 3 January 1915, a Buckinghamshire fruit grower called Charles Burnham sat down with a mug of tea and opened his copy of the *News of the World*. A headline on page three shook him to the core: 'Dead in Bath – Bride's Tragic Fate on Day after Wedding'. The short report explained that one Margaret Lloyd had been found dead at her flat in north London. 'Medical evidence showed that influenza combined with a hot bath might have caused a fainting fit,' it concluded. Charles Burnham's daughter, too, had died in a bath in Blackpool, shortly after her wedding almost exactly a year before. Burnham contacted the police, and discovered that Margaret Lloyd's husband was George Joseph Smith, the man who had previously been married to his daughter, Alice Burnham.

The police called in Spilsbury to perform an autopsy on the exhumed body of Margaret Lloyd. He then travelled to Blackpool to autopsy Alice Burnham. Following this, the police uncovered details of a third woman, Bessie Williams, who had married George Smith and died in very similar circumstances at home in Kent on 13 July 1912.

The coroners had given a verdict of accidental death in the first two drownings but, when the police investigated anew, they discovered that Smith had benefited financially from all of his wives' deaths: he had received £700 and £506 from the life insurance policies held by Margaret and Alice; and £2,500 in trust money (worth around £190,000 today) from Bessie. Now that they could see a pattern, the police arrested Smith.

George Smith and Bessie Williams on their wedding day. She would later become his first victim

From the bodies of Margaret and Alice, Spilsbury could find no signs of violence, poison or heart attack. On Bessie he found evidence of 'goose-skin' on her thigh, which is sometimes an indication of drowning (though it can also be caused by decomposition after death). Spilsbury read the notes of the general practitioner who had first seen Bessie's body, and noted that she had been clutching a bar of soap.

He had all three bath tubs brought to Kentish Town Police Station, where he lined them up together and examined them minutely. He was particularly puzzled by the case of Bessie Williams. Shortly before her death Smith had taken her along to the doctor so she could talk to him about her epilepsy symptoms. Smith had told Bessie that she was suffering from fits, even though she couldn't remember having them and had no epilepsy in her family. Spilsbury was not convinced by this version of events. Bessie was five feet seven inches tall and obese. The bath tub she had died in measured just five feet at its longest, and sloped at the head end. Spilsbury knew that the first phase of an epileptic fit causes complete rigidity of the body and that, given her size and the shape of the bath, such a fit would have raised Bessie's head above the water rather than brought it below.

If an epileptic fit had not been responsible for Bessie's death, what could have caused it? Spilsbury researched further and learned that an extremely sudden rush of water into the nose and throat can inhibit a vital cranial nerve, the vagus nerve, and cause sudden loss of consciousness, swiftly followed by death. A common subsidiary result of this uncommon occurrence is instant rigor mortis – which Spilsbury thought explained the bar of soap clenched in Bessie's fist.

Alfred Swaine Taylor had stated categorically in 1853 that it was impossible to drown an adult without leaving bruises, because of their violent struggle for breath, and this had never been challenged. Detective Inspector Neil decided to carry out a series of experiments before the trial, to test Spilsbury's theory of how the women had died. He found women volunteers who would try to resist being submerged in the bath tubs. The first volunteer stepped into a full bath, clad in a bathing costume, and managed to grip the side of the bath and struggle. But when Neil grasped her ankles and abruptly pulled her legs up, she slid under the water and lost consciousness. It took a doctor several minutes to restore her to consciousness; she was lucky to live. The experiment was not Spilsbury's idea, but he was aware of it and his reputation certainly benefited from its outcome.

George Smith was tried for the murder of Bessie Williams. At the trial Spilsbury spoke with great authority, and effortlessly carried the jury with him. They deliberated for twenty minutes before finding Smith guilty. He was hanged at Maidstone Prison.

Smith had been a silver-tongued conman who had committed his first theft in the East End of London at the age of nine. As he grew up he began to wear gold rings and brightly coloured bowties to help him impress women, so he could exploit them. Because of the early effects of the First World War, and because so many young British men of his day had moved to the colonies, there were a surplus of half a million women in Britain in 1915, providing Smith with plenty of prey. Stoked by the newspapers, public interest in what came to be called the 'Brides in the Bath Murders' was intense at the time. Scores of journalists, hungry for a 'scientist foils serial killer' headline, hounded Spilsbury throughout the investigation. His star would not stop rising in his lifetime.

Many of Spilsbury's cases involved husbands charged with murdering their wives. It's chilling to wonder how many had got away with it before the science had evolved sufficiently to uncover the truth behind those women's deaths. Increasingly, the press presented a picture of Spilsbury as a hero who had devoted his life to interpreting clandestine clues left on defenceless victims, so that their evil killers could not escape justice or strike again. In 1923, that image was bolstered by a knighthood. A year later, another case cemented his reputation ever more firmly.

On 5 December Elsie Cameron left her home in north London to visit her fiancé, Norman Thorne, a chicken farmer in Crowborough, Sussex. They had been engaged for two years, but Thorne had recently started seeing a new girlfriend.

On 15 January 1925, the police found Elsie's dismembered body buried under a chicken run on the farm and her head stuffed into a biscuit tin. Thorne had originally told them that she had never arrived but, after the discovery of the body, he changed his story, saying that she had arrived to tell him she was pregnant and wanted to get married. He said that he had then left the house but, when he returned two hours later, he found her hanging from the ceiling. Assuming it was suicide, he decided to conceal the body, cut it into four pieces and bury it.

Bernard Spilsbury, the eminent and suave pathologist. His testimony helped convict hundreds of criminals, though some of his conclusions have since been challenged

Spilsbury did the autopsy on 17 January and, in his report to the coroner, said Elsie had met a violent death, probably after being beaten. He had found evidence of eight bruises, including one on her temple, not visible on the surface but revealed when he opened the flesh. As he did not see any rope marks on her neck that would indicate hanging, he did not take a section from it. He had noticed two marks on her neck, but thought they were just natural creases. At the inquest, the coroner inquired how it was possible to examine a 6-week-old corpse, but Spilsbury assured him that the decomposition had been no problem. The accused man, Norman Thorne, questioned the report because there were no external signs of bruising, and successfully applied for a second autopsy.

Elsie was exhumed on 24 February and Robert Bronte performed an autopsy with Spilsbury watching on. Post-mortem examinations are supposed to take place in the bright light of day or in electrically lit mortuaries. Elsie was exhumed from midnight to 9 a.m, in front of a crowd of spectators and journalists in the dimly lit chapel at the cemetery. The coffin was full of water, and the corpse had had an extra month to decompose since Spilsbury had examined it, but Bronte saw evidence of marks on the neck and took sections away to analyse.

The trial of Norman Thorne lasted five days. The rival pathologists disagreed over the bruises. When asked by the prosecution whether there were any external marks on the body, Spilsbury replied, 'No, none at all.' The defence pathologist, J. D. Cassels, argued that Elsie Cameron was still alive when Thorne had cut her down from the beam, that the bruises were caused by her falling to the floor, and that she had died from shock ten to fifteen minutes later: this would explain the absence of rope marks, as they were eliminated by the circulation of the blood. He criticised Spilsbury for not examining the neck microscopically.

Spilsbury testified that the two final bruises were those on her face, and that she had been beaten to death with some Indian clubs which were found quite near to the scene. As was his adamant courtroom style, he refused to acknowledge any uncertainty in these conclusions. Yet he had said in a lecture two years previously that if medical evidence is strictly tested by cross-examination, 'that is when the doctor realises his fallibility'.

Throughout the trial, the judge referred to Spilsbury as 'the greatest living pathologist', and told the jury in his summing up that Spilsbury's 'is undoubtedly the very best opinion that can be obtained'. The jury reached its 'guilty' verdict in less than half an hour. Some felt that they had not acknowledged the complexities of the pathologists' evidence, especially the fact that there were no signs that Elsie Cameron had met a violent death. Among those who were concerned with the jury's acceptance of Spilsbury's confident conclusions was Sir Arthur Conan Doyle, who lived close to Norman Thorne. He wrote in the *Law Journal* that 'the more than papal infallibility with which Sir Bernard is readily being invested by juries must tend to be somewhat embarrassing to him'.

Norman Thorne was hanged at Wandsworth Prison for the murder of Elsie Cameron, though he pleaded his innocence to the bitter end. In a famous letter to his father on the eve of his execution he wrote, 'Never mind, Dad, don't worry. I am a martyr to Spilsburyism.'

According to historians Ian Burney and Neil Pemberton, the Thorne trial focused attention on two rival practices of pathology: Spilsbury the celebrity pathologist making a dramatic courtroom appearance, relying on his scalpel and his intuition; and Bronte the laboratory-based pathologist, relying on the latest forensic technology. They argue that Spilsbury's 'virtuosity' in the mortuary and the courtroom 'threatened to undermine the foundations of forensic pathology as a modern and objective specialism'.

In his book *Lethal Witness* (2007), Andrew Rose suggests that Spilsbury caused at least two miscarriages of justice and several more unsafe verdicts. Convictions were sometimes made on flimsy evidence, because if Sir Bernard Spilsbury said a man was guilty then juries believed he must be guilty. In some of his more than 20,000 autopsy reports Spilsbury suppressed evidence because it didn't fit with his narrative.

For example, in 1923, due to Spilbury's evidence, a young soldier called Albert Dearnley was found guilty of tying up his best friend and suffocating him. Convicted of murder, he was only two days away from the gallows when the prison governor read a letter Dearnley had written to a female friend. Its tone worried him and he persuaded the Home Office to grant a stay of execution.

Just in time, the truth came out: the death was not murder, but an accidental asphyxiation that had happened in the course of a sado-masochistic homosexual game. Spilsbury – famous for his homophobia – had suspected the truth, but had kept his counsel because he believed the soldier was a sexual pervert who deserved his fate.

Nevertheless, when Spilsbury committed suicide in 1947, gassing himself in his own laboratory at University College London after a long battle with depression and poor health, it wasn't only the *Lancet* that hailed him as the greatest pathologist of the age. Voices of adulation drowned out the dissenters many times over. His image tarnished only gradually after death. By 1959 his fellow forensic pathologist Sydney Smith was able to write, 'One might almost hope that there will never be another Bernard Spilsbury.'

Today, Dick Shepherd is the UK's best and one of the leading forensic pathologists in the world. But he is adamant that he is no celebrity courtroom performer – and nor would he want to be, even though the roll call of some of his autopsy subjects is remarkable. From Princess Diana and murdered BBC presenter Jill Dando to victims of the 9/11 attacks in the US, he has investigated some of the most notorious deaths of recent years. To him, every case is the same: an autopsy is supposed to be 'a non-judgemental, scientific, acquisition of facts', no matter who the victim is.

What gets Dick Shepherd up in the morning is the living, not the dead. 'I'm fascinated by the interlinking – working with the police, the courts and with other people. Seeing problems, understanding them, interpreting them and providing that information to others. I have to separate myself from the destructive things that I do and remember that I am doing it for the families of the dead. If they understand what has happened it doesn't help them in any practical way, but if there are points of truth they can fix on them, and achieve closure. Forensic science has gone wrong when people haven't been truthful, sometimes because they have hidden the truth in the hope of not distressing families. It never works.'

It's up to the police to make the tricky call about how much a pathologist should know of a case before the first crucial examination of a body. If the pathologist knows too much it might bias the autopsy. If he knows too little he might overlook something important. And, as Dick Shepherd explains, 'If the filtering is done by others they may not give us a critical bit of information. Then when it suddenly pops out in court, it can lead to an "Oh, cor blimey" moment. The lawyer asks, "If you had been told this would you have formed another opinion?" "Yes, I would." "Oh, thank you, Dr Shepherd." Lawyer sits down with smug smile on face.' And the prosecution winces.

One of the reasons Spilsbury so rarely had to witness a defence lawyer's smug smile was that he almost invariably knew a great deal of the background to his cases. Nowadays, when Dick Shepherd gets a call from the police or the coroner's office, he is seldom summoned to the crime scene but rather to the mortuary. Various other scientists with specialisms such as blood spatter and DNA analysis do much of the crime scene evidence gathering that forensic pathologists used to be responsible for. At the scene of death, junior CSIs bag up the body to prevent trace evidence falling from it, such as hair, fibres and dirt, and to protect it from contamination.

When he does go – and 'sometimes it is really useful to, not so much to do any specific examinations but to interpret the scenario' – Dick observes body position and proximity to other evidence such as weapons, fingerprints, points of entry and escape. He has to be very careful not to lose or contaminate evidence by touching or moving the body more than is absolutely necessary. In a recent case of his, the police believed a lady found at the bottom of her stairs had taken a tumble. Dick went along to see 'where she was lying, how she was lying, if she had possibly been moved. When I did the autopsy I found injuries that I thought were caused by her coming down the stairs and banging into things. Because I'd gone to the scene I could later explain the layer of grazes on her side as being where she had gone around a corner.'

Detectives always want an estimated time of death. That's information that can shake, break or confirm a suspect's alibi. The longer a body has been dead, the harder it is to estimate with much accuracy when death occurred; the narrower the estimate, the more useful it is to the investigation.

The first thing a pathologist like Dick Shepherd does when examining a corpse is to take its rectal temperature, unless there is reason to suspect sexual assault. In such cases, he stabs the thermometer into the abdomen. It was once thought that bodies lose heat at a constant 1.8°F per hour until they reach ambient temperature.

So, for example, if a person dies with the average body temperature of 98.6°F in a 68°F room, there will potentially be a 17-hour window when the time of death can be roughly estimated. But research has shown significant variables: a thin body cools faster than a fat one; the larger the surface area to weight ratio and the less subcutaneous fat, the faster the cooling process; whether the body is sprawled or curled up will have an impact; clothing will affect the cooling; shade or sunlight; in the shallows or beside a river. Even so, an early and careful reading can be a useful starting point and pathologists can factor in variables like ambient temperature and body weight using a graph with multiple axes called a nomogram.

The next phenomenon that interests Dick is rigor mortis – the macabre reason why dead bodies are known as 'stiffs'. The symptoms of rigor are useful to the pathologist for about two days after death because of a recognised cycle. At first a body relaxes completely, then after three or four hours the small muscles of the eyelids, face and neck begin to stiffen. Rigor progresses downwards, from head to toe towards the larger muscles. After twelve hours the body is completely rigid and will stay fixed in the position of death for around twenty-four hours. Then the muscles gradually relax and stiffness goes away in the order in which it appeared, starting with the smaller muscles and progressing to the larger ones. After a further twelve hours or so all the muscles reach a state of complete relaxation.

But even a process as well documented as rigor mortis is a very imperfect indicator of time of death. The hotter the ambient temperature, the faster each step of the cycle occurs. Also, bending and stretching a corpse breaks up its muscle fibres and eliminates rigor, a piece of information that murderers have been known to employ to confound the investigation.

Rigor mortis is followed by the least majestic stage of the body's time on earth. 'Putrefaction' may not be a loveable phenomenon, but forensic pathologists need to be on intimate terms with it to do their jobs properly. First the skin around the abdomen turns greenish as the bacteria in the lower intestine begin a process of 'self-digestion'. As bacteria grow throughout the body, breaking down proteins into amino acids, they produce gases and cause the body to bloat – starting with the features of the face, where the eyes and tongue begin to protrude. Next, a web-like pattern of blood vessels appears as marbling on the skin because the red blood cells break down and release haemoglobin. Gases continue to swell the abdomen until they escape, sometimes explosively, and produce a vile smell. The body has turned greenish-black by now, as fluids drain from the nose and mouth, and the skin falls away like 'a giant rotting tomato'.

All the while, and aided by 'self-digestion', the internal organs have been liquefying, starting with the digestive organs and the lungs, and then the brain. Flies have laid eggs on the body's points of entry, such as the eyes, mouth and open wounds, and maggots have been tearing continuously at the flesh.

Scientists are continually studying and refining the various different ways of measuring time since death. However, as forensic anthropologist Sue Black explains, this doesn't always simplify matters. 'The more information we get, the more we realise just how difficult it is. No two bodies will decompose in the same way, and at the same rate. You can have two bodies that are literally six feet apart and they will decompose in entirely different manners. It could be the amount of fat on the body. It could be the drugs they were taking, or the medication. It could be the type of clothing they're wearing. It could be that one has a particular odour that is more attractive to flies than the other. Absolutely anything.'

One way to try and combat the headache of variables is to develop new tools. That's what the Anthropological Research Facility at the University of Tennessee has been doing for years. 'The Body Farm', as it's better known, was set up in 1981 by William Bass to research putrefaction. It was the first institution to develop the systematic study of human decomposition and how bodies interact with the environment. More than 100 people donate their bodies to the Body Farm every year, to be placed in different settings and left to rot. Researchers have worked out a general rule of thumb: one week exposed above ground equals eight weeks below and two weeks in water.

Arpad Vass, Associate Professor of Forensic Anthropology at the University of Tennessee, is developing a new method for estimating time since death. 'Decomposition Odour Analysis' hopes to identify the 400 or so distinct vapours that a body gives off at different stages of its decomposition. Understanding when these vapours are given off, in a variety of settings, and measuring them in a corpse, could provide a more accurate time of death than has so far been possible.

Findings from research facilities such as the Body Farm trickle into the world of practical forensics through journals and monographs, and arm pathologists with knowledge that they can use to give better evidence to criminal investigations. Pathologists use that knowledge most usually in a morgue or a hospital, locked in the intense focus of autopsy. How and why did the person die? Was it suicide, murder, an accident, old age, or is it impossible to tell? There are seldom straightforward answers. A bullet that has passed through someone's head could have been shot suicidally, murderously or accidentally. The scope of a forensic pathologist's curiosity is very broad when he enters the morgue. It gradually narrows its focus to the small details, before broadening out again to incorporate those details in a conclusion. The general steps of an autopsy have changed little since the beginning of last century.

When the body arrives in the morgue Dick Shepherd is ready to photograph it. An assistant takes the bag the body was stored in for transportation and searches it for trace evidence. Dick removes the subject's clothes, which he photographs, bags and documents. Then he takes biological samples, plucking hair, scraping fingernails, swabbing sexual organs. Only now does he carefully take the fingerprints; prying open fingers to take prints can jeopardise small bits of trace evidence clenched in a fist closed by rigor.

Dick then washes the body and documents every scar, birthmark, tattoo and unusual physical feature he can find. 'Every pathologist has a different route,' says Dick. 'I start at the head and I always do the left-hand side first. So I do head, chest, abdomen, back, left hand, right hand, left leg, right leg. As I go around, injuries are all documented and photographed. I have to say my heart sinks now when it's a pub fight and there's 970 2 cm bruises. Can't I just say there were lots of bruises on the legs? No.' In less clear-cut cases the rigorous style encouraged by the documenting process can prove invaluable, as illustrated by another British wife who died in a bath.

At 11 p.m. on 3 May 1957, Kenneth Barlow, a nurse from Bradford, phoned 999 to say that he had found his wife unconscious in the bath. He explained that he had pulled her out and spent a long time trying to resuscitate her, and that she had been suffering from vomiting and fever that evening. Investigators were suspicious when they discovered two used syringes in the kitchen. Kenneth explained that he was using them to treat an abscess he had with penicillin. Tests confirmed the presence of penicillin.

But pathologist David Price remained suspicious. During the autopsy he searched every inch of Mrs Barlow's skin with a magnifying glass. Eventually he found two tiny holes consistent with injection needles, one on each of Mrs Barlow's buttocks. The symptoms that Kenneth had said his wife was suffering from were those of hypoglycaemia (low blood sugar), which made David Price suspect that he had injected his wife with a lethal dose of insulin. There were no tests for insulin at the time, so Price took tissue from around the injection points on Mrs Barlow's buttocks and injected it into mice. They quickly died of hypoglycaemia. Barlow was found guilty of murder and given a life sentence.

After the meticulous external scrutiny, the internal examination begins. The pathologist is looking both for internal injuries and any medical condition that might have caused the person to die naturally. Dick Shepherd cuts open the body in a 'Y' shape from both shoulders down to the groin, sawing open the ribs and collarbone and removing the breastplate, to reveal the heart and lungs. He inspects the neck, looking out for things like broken cartilage, which might indicate strangulation. He then takes out the organs individually (like the liver) or in groups (like the heart and lungs), examines their surfaces, and makes cuts to examine them internally. He preserves samples of the organs. 'The Home Office [the UK government department responsible for national policing policy] now insist that we do microscopic examinations of all of the major organs on every case, even when it's an 18-year-old who's been hit by a baseball bat on the top of the head.' Better safe than sorry, for which we should all be grateful. He then sends those samples off to the laboratory. Next Dick makes an ear-to-ear incision over the top of the head and peels back the scalp. Now he can saw away a section of the skull and look at the brain in place before removing it for a closer examination.

Finally, Dick sews up all the incisions he has made in the organs, places them carefully back in the body, and stitches up his initial 'Y' incision. Afterwards he talks to detectives and to other forensics experts, pools ideas about what looks suspicious or needs to be followed up, and feeds back into the inquiry. Very often there will be a second autopsy, so that another pathologist can check Dick's findings. Once all the reports have come in from the experts – bone pathologists, neuropathologists, paediatric radiologists – Dick writes his report for the coroner.

In some extreme situations more than one autopsy will follow Dick's. On 23 August 2010, the police found a red North Face bag in the bath of a flat in Pimlico, London. It belonged to Gareth Williams, a Welsh maths prodigy who had been working as a code-breaker for MI6, and its zips were fastened with a padlock on the outside. When the police prised the padlock open they found the naked, decomposing body of 31-year-old Gareth folded inside.

The police regarded the death as suspicious. Gareth Williams' family were convinced that M16 or another secret service was involved in the death, as Williams had been working with the FBI as part of a team attempting to penetrate hacking networks.

Dick Shepherd was one of three pathologists who performed autopsies, and who all agreed that Williams had been dead for about seven days. Whilst they found no signs of strangulation or physical trauma, it was difficult for them to determine the cause of death because the body had decomposed so quickly: the bag had been found in the summer, and all the radiators in the house had been left on maximum. Toxicologists found no signs of poisoning but, because of the body's condition, they couldn't rule it out. Suffocation seemed most likely.

The autopsies revealed small abrasions on the tips of Williams' elbows, which might have resulted from him moving his arms inside the bag, possibly in an attempt to escape. Dick Shepherd summed up the situation: 'Once the lock was placed on it there was no possibility of getting out of that bag. The question is: did he place the lock there or did another?'

Peter Faulding, an ex-military reservist who specialised in rescuing people from confined spaces, told the inquest in May 2001 that he had tried 300 times to lock himself into an identical 81 cm by 48 cm North Face bag, and failed to do so. He said that even Harry Houdini 'would have struggled to lock himself in the bag'. Another expert had tried and failed a further 100 times.

But Dick Shepherd felt that Williams had suffocated, and was 'more likely than not' alive when he entered the bag. He reasoned that, because Williams was hunched into such a tight ball when he was found, and because of how 'floppy' a body is before rigor mortis sets in, it would have been very tricky for someone else to stuff him inside it. No expert offered to test this theory. It was also revealed at the inquest that DNA found on Williams' finger during his first autopsy did not come from a mysterious Mediterranean couple, as initial DNA results analysis had suggested and whom the police had been pursuing for a year. An employee at LGC, the forensic company which was analysing samples, had typed the wrong DNA details into a database. The DNA in fact belonged to one of the CSIs who had been at the scene. LGC expressed 'very deep regret' to the Williams family for the mistake.

Twenty thousand pounds-worth of designer women's clothes had been found at Williams' flat, along with women's shoes and wigs. Investigators had also found pictures of drag queens and evidence that he had been browsing self-bondage websites and sites about claustrophilia – the love of enclosure – days before his death.

The coroner, Fiona Wilcox, ruled that, though there was not enough evidence to give a verdict of unlawful killing, it was probable that the death was unnatural, that someone else had locked him in the bag and placed it in the bath, and that he was alive when he entered it. She added that there was no evidence to suggest that Williams was a transvestite 'or interested in any such thing'.

Days after the verdict, a 16-year-old girl and a 23-year-old journalist separately attempted to lock themselves into identical North Face bags: they climbed in, drew their legs up, pulled the zippers nearly shut, poked their fingers through the gap, and closed the padlock. They then tensed their bodies, and the zip sealed itself up. The journalist, who was of a similar build to Gareth Williams, repeated the trick a number of times until she could do it in three minutes.

Peter Faulding didn't think much of these stunts. 'None of my conclusions are wrong. A young girl zipping a bag doesn't discredit this inquiry whatsoever. We were fully aware of other methods of being able to lock the bag but she or nobody could achieve it without leaving her DNA or trace on the bath and that's the key to this.'

Dick Shepherd continues to disagree. 'I'll never convince the coroner. She was furious with me; the word "incandescent" springs to mind. One of the strongest things that makes me feel he was on his own is that he lived a solitary existence, wearing women's clothes, working intensely, and being a mathematics geek – forget the pathology, the psychology is not right.'

In 2013, the Met Police held an internal investigation because the coroner had indicated that MI6 staff might have been involved. By November Scotland Yard had the investigation's verdict: it was probable that Williams had died alone, as a result of accidentally locking himself inside the bag. Dick Shepherd's view had been vindicated.

◆ ◆ ◆

Mysterious cases often need imagination to bring them to a resolution: the pathologist who injected tissue from a woman's buttocks into mice; the journalist who locked herself into a small bag when a military expert couldn't do it. These people wanted to follow the original Greek meaning of autopsy, to see for themselves. Our curiosity grows with every fresh technique we road-test. New technology is letting pathologists see deeper than ever before into a human subject, without the need to roll their sleeves up. The Virtual Autopsy (VA) table is a new medical visualisation tool built in Switzerland that combines CT and MRI scans to transform images of a dead body into a 3D computer model. Pathologists using it in Germany have found fractures and haemorrhages undiscoverable using conventional autopsies. The VA table also includes a high-resolution scanner, which means the skin can be magnified to better reveal things like bruising or malicious injection points. It also causes less distress to the living, who don't want to contemplate what they see as the desecration of the bodies of their loved ones.

Some traditional forensic scientists have called the VA table unproven and new-fangled. But as a younger, more tech-savvy generation move into pathology labs, they have begun installing them. By January 2013 three out of the thirty-five forensic science institutes at German universities had one. Forensic pathologists are tending still to use them as a complement to physical autopsies. But the proofs keep building up. In the case of a mountain climber who had fallen to his death in the Swiss Alps, his shattered neurocranium, fractured lumbar spine and broken lower leg were all detected without a single scratch from a knife.

Another benefit of the Virtual Autopsy table is that the 3D model it produces can easily be examined by several pathologists independently, saved for future reference and produced in court for juries to judge for themselves. Spilsbury might not have liked the idea, but his martyrs surely would have.

FIVE

TOXICOLOGY

'Within the infant rind of this weak flower
Poison hath residence and medicine power'

Romeo and Juliet, II, iii

Drugs are frighteningly ambiguous. A small amount of digitalis from the foxglove plant smooths abnormal heart rhythms. But too much can provoke nausea and vomiting, and send the heart careering catastrophically towards death. Paracelsus, the founder of modern toxicology, expressed this idea neatly when he wrote in 1538, 'The dose makes the poison.'

Poison is one of the oldest weapons humans have used against each other. As science advanced, the job of the toxicologist developed to identify lethal substances and search for antidotes. One man in particular systematised the field. Mathieu Orfila studied in Valencia and Barcelona in the early nineteenth century before moving to Paris to study medicine. In order to find out their effects, Orfila spent three years testing poisons on several thousand dogs, who suffered horrendously. (Anaesthetics weren't available until the 1840s, and besides, they would have contaminated his experiments.) At the age of just twenty-six he published his encyclopaedic *General System of Toxicology; or, A Treatise on Poisons* (c.1813), which catalogued all the known mineral, vegetable and animal poisons. The 1,300-page work remained the principal reference work on toxicology for forty years.

In a key section of the *Treatise,* Orfila described his improvements to the existing tests for the substance synonymous with our image of the nineteenth-century poisoner – arsenic. Orfila had realised that severe vomiting can remove all traces of arsenic from a person's stomach. By testing the organs of his poisoned dogs, he learnt that the bloodstream spreads arsenic throughout the body. He also demonstrated that buried bodies could absorb arsenic from the soil surrounding them, which can make it appear as though they were poisoned in life. After the *Treatise* toxicologists tested the nearby soil whenever a body was exhumed.

In 1818 Orfila published *Directions for the treatment of persons who have taken poison; together with the means of detecting poisons and adulterations in wine; also, of distinguishing real from apparent death,* in order to 'render popular the most important information contained in my Treatise on Poisons'. People were coming to realise that appropriate first aid could limit the harm of an accidental poisoning. Orfila was both genuinely concerned about the ignorance of the general public, and aware that there was money to be made from this new scientific field. He wrote in the introduction to his book, 'It is of the highest importance that the clergy, the magistracy, the heads of large establishments, the fathers of families, and the inhabitants of the country' inform themselves about toxicology. Numerous translations of the book into German, Spanish, Italian, Danish, Portuguese and English cemented Orfila's reputation. If a lawyer needed a toxicologist to testify at a trial, Orfila was the one to get, especially after he became royal physician to Louis XVIII.

In 1840 Orfila became involved in a cause célèbre, the murder trial of the delicate and refined heiress Marie-Fortunée Lafarge. People came from all over Europe to see her fate decided.

Marie had grown up as an aristocrat in Paris, watching on as her school friends married wealthy men. By the age of twenty-three her desire to do the same had reached such a pitch that her uncle employed a professional marriage broker. It was an easy job.

After all, Marie had youth, beauty and a dowry of 100,000 francs. The broker contacted a bachelor, Charles Lafarge, the owner of a thirteenth-century monastery in the Limousin region of central France.

The buildings on the Lafarge estate had started to crumble, but Charles had been determined to restore the family fortunes. He set up a forge in which he invented new smelting techniques. He poured his money into the enterprise, but it stuttered and eventually he was forced to shut down the furnaces. By 1839 he was nearing bankruptcy. There seemed only one route to salvation – a woman with money. Charles had contacted the distant Parisian marriage broker, leaving mention of money troubles out of his profile and concentrating instead on a 200,000 franc evaluation of his estate and a glowing letter of reference from his priest.

Marie took an immediate dislike to Charles. She found him boorish and wrote in her diary that 'his face and figure were most industrial'. But she did like the idea of an extensive estate, sumptuous couches to lounge on, fragrant gardens to walk in. And surely the owner of an ancient monastery would have a hidden poetry in his soul?

Within four days of meeting the pair were married and sharing a carriage down to Limousin. When Charles began to eat a roast chicken with his bare hands, washed down with a whole bottle of Bordeaux, Marie chose to sit in front with the driver. When they arrived at the house she got an even bigger shock. Her in-laws were dressed 'in the worst provincial taste', the furniture was 'shabby and ridiculously old-fashioned', and the place was full of rats. On that first night, 13 August 1839, Marie locked herself in her room and wrote an impassioned letter to her husband, begging him to release her from the marriage, 'Or I will take arsenic, for I have it on me … I am willing to give you my life. But receive your embraces, never.'

After she had calmed down, Marie agreed to stay with Charles on one condition. He could not consummate the marriage until he had secured enough money to renovate the estate. It seemed to the other members of the household that the couple were getting along rather better. Marie enjoyed walking around the ruins of the Gothic church and cloisters. She wrote letters to her school friends which depicted a scene of domestic bliss. She didn't mention having to buy arsenic to keep down the vermin, however.

Marie then suggested to her husband that she would write a will leaving all she owned to him, on the condition that he do the same for her – standard behaviour for a new couple in love. But Charles cunningly made a secret second will leaving everything to his mother.

Four months after the wedding, Charles returned to Paris for a Christmas business trip to raise funds. While he was away Marie sent him tender love letters expressing how much she missed him, and a homemade Christmas cake. When Charles ate a piece, he vomited soon afterwards. He returned to Limousin, having raised some money but still feeling nauseous. Marie received him with concern and suggested the only place for him was bed. There she fed him truffles and venison. But his condition deteriorated and the family doctor was called. He feared cholera, which put the household in a state of panic.

The next morning Charles had acute cramps in his legs and terrible diarrhoea. No matter how much water he drank he couldn't keep it down. A second doctor was called, and he agreed that Charles had cholera and suggested that he drink eggnog to build up his strength. But Anna, one of the women employed by the family to tend Charles, noticed Marie stir some white powder into the eggnog before giving it to Charles. When she asked Marie about it she said it was 'orange-blossom sugar', but Anna was suspicious, and she hid the eggnog away in a cupboard.

On the afternoon of 13 January 1840 Charles Lafarge died. By this point Anna had told other members of the household of her fears. Marie's calm reaction to the death of her husband, which had at first seemed dignified, began to look suspicious. The next day Marie went to her notary with what she believed was Charles's last will and testament.

Meanwhile, Charles's brother went to the police. Two days after Charles's death, a justice of the peace came to the estate, arrested

Marie and launched an investigation. Local doctors tested Charles's eggnog, his stomach and some of his vomit. They found traces of arsenic in the eggnog and stomach, but nothing in the vomit.

Things looked dark for Marie, but her lawyer had an idea. 'Knowing that in such affairs M. Orfila is the prince of science', he wrote to him. Orfila replied explaining that the local doctors had used arsenic tests dating back to the seventeenth century. What was needed was his refined version of a test developed four years ago by the English chemist James Marsh. When Marsh published the details of his extremely sensitive test for arsenic the *Pharmaceutical Journal* of London enthused: 'The dead are now become the witnesses whom poisoners have most to fear.' There were some problems with the Marsh Test, but Orfila had worked through most of these. Backed up by another test invented two years later by Hugo Reinsch, the Marsh Test would remain the standard for arsenic until the 1970s, when more sophisticated methods were developed using gas chromatography and spectroscopy. Armed with Orfila's letter, the lawyer discredited the original test, and the judge ordered the local doctors to do it again according to the more modern Orfila version.

The doctors performed the test on Charles Lafarge's stomach, his vomit and the eggnog. This time, they found nothing.

By now the prosecuting lawyer had got hold of a copy of Orfila's *Treatise on Poisons* and read it carefully. Severe vomiting, he now knew, could remove all traces of arsenic from somebody's stomach. Moreover, once the bloodstream passed through the stomach it could take arsenic to the other organs. He told the judge it was necessary to dig up Charles's body and test his organs. The judge agreed and the local doctors performed another Marsh test, this time in front of a crowd of spectators, some of whom fainted at the 'fetid exhalations'. Yet again, they found no arsenic. Upon hearing the news in court, Madame Lafarge wept tears of joy.

In a last ditch effort, the prosecutor asked the local doctors how many times they had carried out the Marsh Test in their career. Not once, they admitted. The prosecutor pleaded with the

Marie Lafarge, who was convicted of murdering her husband Charles with eggnog laced with arsenic

judge that this trial was too important to be decided by a couple of provincial doctors. The only person fit for the job, he argued, was the world's leading toxicologist, Doctor Mathieu Orfila himself. Arriving on the express train, Orfila set to work immediately, macerating what was left of the organs, 'liver, a portion of the heart, a certain quantity of the intestinal canal, and a part of the brain'. And this time, the Marsh Test *à l'*Orfila produced a positive result. For good measure, Orfila demonstrated that the arsenic had not come from the soil by Charles's coffin.

Madame Lafarge was sentenced to life imprisonment with hard labour. From jail she published her memoirs in 1841, protesting her innocence, which she continued to do until she died of tuberculosis aged thirty-six.

Orfila's performance of the Marsh Test would come to be seen as a watershed moment in the fight against murder by poison – a vindication of forensic toxicology. Yet in the immediate aftermath of the trial the public felt dizzy, unable to decide if forensic toxicology was a science, an art or a game. One newspaper summed it up: 'Within two days the accused was declared innocent by the verdict of science, and now she is judged guilty by the verdict of that same science.' Getting a forensic toxicologist involved in a suspicious murder appeared to be only half the battle. It had to be the right forensic toxicologist.

Marie Lafarge was one nineteenth-century arsenic murderer among many. Her fellow poisoners were motivated by money, revenge, self-defence, even sadism. The French made their most popular motivation clear when they dubbed arsenic '*poudre de succession*' (inheritance powder). On the other side of the Channel, in England and Wales, there were ninety-eight criminal poisoning trials between 1840 and 1850, and *The New York World* trumpeted in 1899 'Poison Epidemic Sweeps the Land.' It might seem strange that such a poisonous decade came on the heels of the 1838 Marsh Test. But the truth is that, before the test, coroners were much more likely to pronounce victims of arsenic 'dead by natural causes'.

The reason was the difficulty of establishing it as a murder method. Arsenic was virtually tasteless – some even said it had a slightly sweetish taste – odourless, and cheaply available from all manner of shops. The body cannot excrete it, so the heavy, metallic element builds up in the victim's system, mimicking the slow deterioration of a natural disease. Those who digest it suffer a range of symptoms with varying degrees of severity. Hypersalivation, abdominal pain, vomiting, diarrhoea, dehydration and jaundice can all be a result of arsenical poisoning. Because reactions are so variable, killers could strike more than once without arousing the suspicion of local doctors, who diagnosed an apparently random mixture of cholera, dysentery and gastric fever. Intelligent arsenic killers usually went down the chronic rather than the acute path of administration, favouring introducing small amounts over a long period rather than one large dose, as sudden, violent death throes encouraged suspicion.

In response to this, in 1851 Parliament passed the Arsenic Act, making it harder to buy arsenic over the counter. Sellers had to be registered and buyers had to sign and give a reason for their purchase. Unless it was for medical or agricultural use, all arsenic had to be coloured with soot or indigo, so it looked less like sugar or flour.

But the Arsenic Act and the Marsh Test did not deter all would-be killers. In 1832, Mary Ann Cotton (neé Robinson) was born in a village near Durham in the north-east of England. When she was nine, her father died after falling down a mine shaft and the family fell on hard times. Mary Ann was a bright girl and in her teens she taught at the local Methodist Sunday School.

When she was nineteen Mary Ann fell pregnant to a miner called William Mowbray and together they travelled the country looking for work. Mary Ann gave birth to five children during this nomadic period, but four of them died, possibly from natural causes.

In 1856 the couple moved back up north, where Mary Ann had three more children by Mowbray, all of whom died of diarrhoea. Her grief didn't prevent her from claiming on the life insurance policies she'd taken out on all three of them. Then Mowbray injured his foot in a pit accident and had to convalesce at home. Soon he became ill and was diagnosed with 'gastric fever', and died in January 1865. Mary Ann went down to the office of the Prudential Insurance Company and collected the £30 policy which she had recently encouraged Mowbray to take out.

Over the next dozen years Mary Ann became the most prolific female serial killer in British history. Although it will never be known exactly how many people she poisoned with arsenic, she likely murdered her mother, three of her four husbands (the other one refused to take out a life insurance policy), a lover, eight of her twelve children, and seven stepchildren – at least twenty people in total.

In 1872, Mary Ann set her sights on Richard Quick-Mann, a customs and excise officer who was significantly richer than her previous working-class husbands. Only her 7-year-old stepson, Charles Cotton, stood in the way. She tried fostering him with one of his uncles but failed. Then she took him to the local workhouse.

When the superintendent refused him entry unless Mary Ann accompanied him into their care, she told him that the boy was sickly and if the superintendent didn't change his mind, he would soon die 'like all the other Cottons'.

All other options having failed, she poisoned Charles. The workhouse superintendent heard about his sudden demise and went to the police. The doctor who had attended Charles before he died carried out an autopsy and found no evidence of poison. So the coroner ruled death by natural causes. But the doctor had kept Charles's stomach and intestines and, when he subsequently performed a Reinsch Test on them, he discovered the lethal poison.

The bodies of Mary Ann's most recent victims were exhumed and found to contain high levels of arsenic. Her defence lawyer argued that Charles had inhaled arsenic fumes from green paint on the wallpaper of his room. But, under the weight of the evidence from the exhumed bodies and other witness statements, Mary Ann was found guilty of murder and sentenced to death. It seems extraordinary now that no one suspected her, but until she poisoned Charles she had been too careful, clever and charming, changing her name and moving house too often, to be found out. Moreover she lived in a world in which infant mortality among the working classes may have been as high as 50 per cent.

But once she was hanged, her notoriety was guaranteed. A broadsheet rhyme was coined that began 'Mary Ann Cotton – she's dead and she's rotten', and the story ran in the newspapers for months. Did she just do it for the money? Or were there even darker forces at play? Could it happen again? Why did it take so long to catch her? Might anyone be able to get away with a poisoning?

The Victorians were fascinated by the figure of the female poisoner, oozing loveliness and sweetness, offering her husband a second spoonful of sugar for his tea and then making it a lethal one. Readers found a mixture of fascination, fear and excitement in this literally femme fatale image. In fact, more than 90 per cent of convicted spouse murderers in nineteenth-century Britain were men. But men were far more likely to stab or strangle their wives; twice as many wives as husbands stood trial for the more indirect murder method of poisoning.

It wasn't always a straightforward case. Arsenic abounded in everyday life. Arsenical paint found its way on to children's toys, nursery book covers, green wallpaper and curtains; cosmeticians incorporated it into beauty products; it was an ingredient in virility pills, pimple creams and cheap beer. As a result, in cases of unexpected death, toxicologists had to be sensitive to the amount of arsenic in a corpse, so as to avoid wrongly charging someone with murder.

Down the years manufacturers have used a variety of other poisonous elements in their products, sometimes in ignorance of their ill effects, sometimes hoping to keep everyone else in ignorance. In the early twentieth century the work of two New York doctors would have lasting implications for negligent corporations as well as for would-be killers.

In 1918 Charles Norris set up the first organised medical examiner system in the world when he became New York City's first Chief Medical Examiner, responsible for investigating the bodies of people who had died unnaturally or suspiciously. Previously, forensic pathology had been the preserve of 'elected coroners', who were generally barbers or undertakers or worse. Forensic historian Jurgen Thornwald counted 'eight undertakers, seven professional politicians, six real-estate agents, two barbers, one butcher, one milkman [and] two saloon proprietors' serving as elected coroners in New York between 1898 and 1915. The system was incompetent and corrupt. But now the Chief Medical Examiner and his staff would have to be doctors of medicine as well as 'skilled pathologists and microscopists'.

Norris appointed Alexander Gettler as Pathological Chemist, and asked him to set up the first forensic toxicology lab in the US. Gettler set about inventing a host of techniques for uncovering toxins. At a time when poisonings from bootlegged alcohol were reaching epidemic levels, Gettler developed many new ways to identify the active ingredients. Each time he dealt with a case involving an unknown toxin he took a piece of liver from the local butcher's shop, injected it with the toxin and experimented until he could recover and identify it.

He studied more than 6,000 brains to come up with 'the first scientific scale of intoxication'. After Gettler, pathologists started testing brain tissue for the presence of alcohol in all violent or unexplained deaths. He also devised tests for chloroform, carbon monoxide, cyanide, blood and semen, among other substances. So when science itself was put in the dock, Norris and Gettler were the obvious experts to put it under scrutiny.

The story begins in Paris in 1898, with Marie Curie's discovery of a trio of radioactive elements, thorium, polonium and radium, and the subsequent exploitation of their properties. By 1904 doctors had started using radium salts to shrink cancerous tumours, which they called 'radium therapy'. It was seen as the new miracle substance – radium water, radium soda, radium facial creams, radium face powder, radium soaps were all the rage. Advertising hoardings were full of the glowing element, rejuvenator of body and soul.

Nothing seemed beyond radium's benevolent rays. The US Radium Corporation even applied radium paint to watch faces to make them emit a pale greenish glow. By the end of the First World War, glow-in-the-dark watches had found their way on to the wrists of fashionistas across the United States, and the Radium Corporation was doing a roaring trade.

The dial painters at the Corporation's factory in Orange, New Jersey, painted around 250 watch faces a day. Their managers instructed them to be as neat as possible when applying the expensive paint to the watches; they were taught to make the brush tips come to a sharp point with their lips. These were young

A contemporary advert for a radium-based facial cream, 'made to the formula of Dr Alfred Curie'

women, and when they had a break they used to paint their fingernails and streak their hair with the radium paint; one of them even gave herself a spooky smile by covering her teeth.

But by 1924 the Orange dial painters had started to fall ill. Their jawbones were rotting. They lost the ability to walk as their hips dislocated and their ankles cracked. They were constantly tired from low levels of red blood cells. Nine died. Worried about the business repercussions, US Radium hired a group of scientists from Harvard University to investigate. They concluded that the deaths were 'connected' to the factory work. The nervous management were so scared about the profit implications that they prevented the report from being published. But another group of scientists also carried out tests on the workers.

Forensic pathologist Harrison Martland read their report and was determined to investigate further. Martland was a passionate

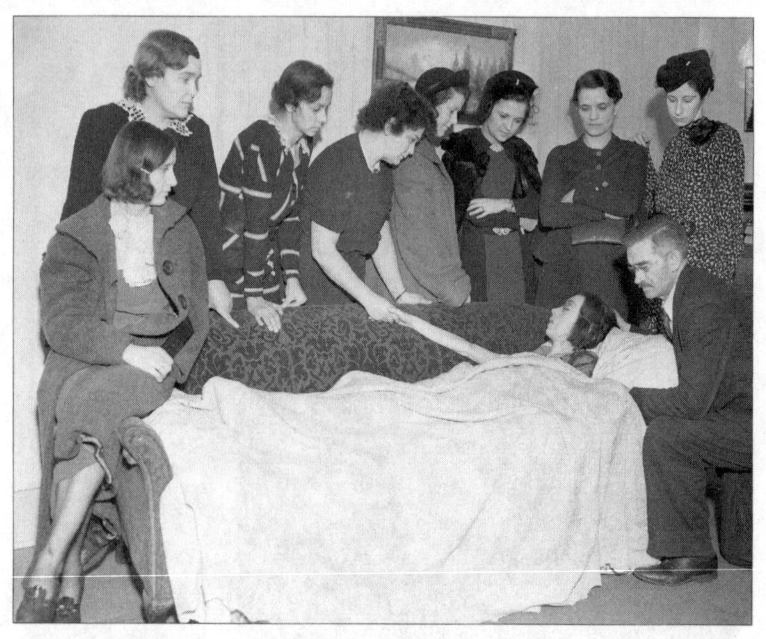

Nine of the 'Radium Girls', whose jobs painting watch faces with glow-in-the-dark paint gave them fatal radiation poisoning

campaigner for workplace safety, publishing research demonstrating that nitroglycerine was poisoning workers in explosive factories, and that beryllium, used in the fledgling electronics industry, could cause fatal lung diseases. Regulation on both those chemicals soon followed his work.

Martland studied the bodies of living and recently dead Orange workers, and published his findings in 1925. The element radium is structurally related to calcium, he explained. When it is ingested, the body treats it like calcium: some is metabolised, some is transmitted to the nerves and muscles, most is deposited in the bones. But unlike calcium, which strengthens bones, radium bombards them with radiation, destroying the blood-building marrow at their core and creating tiny holes, which get larger with time.

That year a small group of former employees took the brave step of suing US Radium. It took the 'Radium Girls' – as the press quickly dubbed them – three years of legal wrangling just to get a trial date.

Meanwhile, Martland had asked Charles Norris at the New York Medical Examiner's Office to gather evidence for the trial. Together they planned to exhume the body of former dial painter, Amelia Maggia, who had died at the age of twenty-five. In her last year of work she had lost weight and suffered joint pains. The following year her jaw had started splintering, and nearly all of it had had to be removed. She had died in September 1923, of 'ulcerative stomachitis', in the words of the coroner.

Norris asked Alexander Gettler to analyse Amelia's bones, including the skull, feet and right tibia. Gettler's team boiled them for three hours in a solution of washing soda. Then they sawed the larger ones into two-inch pieces. Gettler brought the bones into a darkroom which contained X-ray films. He sealed the bones tight up next to the X-ray films and did the same for some control bones from another corpse. When Gettler came back to see the results ten days later, the X-ray films around Amelia Maggia's bones had a dazzle of pale spots on them, and the control films had nothing. He published the results of his experiment.

As the lawsuit dragged on, the condition of the Radium Girls deteriorated. Two of the five girls were Quinta and Albina Maggia, sisters of Amelia. Both of Quinta's hips had fractured and Albina could not leave her bed; by now, one of her legs was four inches shorter than the other. Another woman, Katherine Schaub, was hoping to use the money to buy roses for her funeral.

The defence lawyers for US Radium tried to stall things further, arguing that the women couldn't sue because they weren't working at the factory any more. But the prosecution drew on the research of Martland and Gettler to argue that, while traditional toxins like arsenic and mercury poison you for a period of time, radium stays with you for ever. When the Radium Girls breathed out, all five of them exhaled radon gas.

The courts dismissed US Radium's motion and insisted the trial go ahead. This prompted them to settle, giving each of the women $10,000 cash, annual pensions and free health care. It was a cheap settlement; at least two of them were dead within the year.

The sad story of the Radium Girls is told by Deborah Blum in her book *The Poisoner's Handbook* (2010). The length of time it took for the employers to be punished and the victims to be given some kind of justice speaks to the modern problem of industrial worker poisonings. James C. Whorton, author of *The Arsenic Century* (2010), has written: 'As with arsenical candles and papers and fabrics, items become established in commerce before their dangers are recognized, ensuring that any attempt to curtail their use will be resisted by manufacturers ... and fought or ignored by politicians ideologically opposed to government interference ...'

Gettler's forensic toxicology laboratory became a model for others. The combined endeavours of scientists narrowed down the list of untraceable toxins until virtually none remained. But although the use of poison as a murder weapon has tailed off, and working conditions for industrial workers have improved in developed countries, the number of people injured or killed from 'drugs of abuse' – heroin, cocaine, crystal meth – remains large. This is the area in which forensic toxicologists have become most heavily involved lately.

Robert Forrest is Honorary Professor of Forensic Chemistry at the University of Sheffield, and Britain's leading authority on forensic toxicology. His route into forensics started when he set up a clinical toxicology service in Sheffield, enlisting a range of high-tech instruments in the service of state-of-the-art analysis. Among their other work, Robert and his team started analysing post-mortem samples following a spate of deaths from the heroin substitute methadone.

Then the local coroner contacted Robert to ask if he would help with forensic investigations. 'And that of course would bring a bit of money in, so I started doing that and it grew from there,' he says. The work was new and difficult and Robert's expertise grew.

Because most toxins don't make a visible difference to bodily tissues, even under the microscope, Robert needed to chemically test samples of blood, urine, organs, hair and, more recently, toenails supplied by the pathologist.

Sometimes methadone poisoning is chronic rather than acute, and Robert can reveal that from the victim's hair. Hair grows at about a centimetre a month, so Robert chops a hair sample into centimetre segments and analyses each one to provide a timeline of drug ingestion. The technique is useful in drug screening and also in what is known as drug-facilitated assault. 'The sort of case where it's useful is when there is a prostitute who has got a kid who she needs to keep quiet while she is entertaining clients and she feeds the kid a smidge of methadone and then one day she feeds him too much. Her defence is that somebody else must have fed it to him, but then you find in the hair lots and lots of methadone ingested over a period of several months, which makes that defence not terribly good.'

However, this is not a foolproof method. For example, light hair binds drugs less well than dark hair because it contains less melanin. And cosmetic treatments such as dyeing and straightening tend to strip out drug traces. Nevertheless hair remains a useful indicator of drug content, not least because it remains stable after death.

Something that became clear to Robert over time was that drug concentrations change significantly after death in most other parts of the body. 'Interpreting the results is not at all straightforward,' he admits. The scientific consensus used to be that 'living blood' gives the same toxicological results as post-mortem blood. 'Nowadays we know that's not true. You have to look at it with a great deal of caution. It's very, very difficult.'

How much toxin and where it can be found in the body depends on how it was taken. If it was inhaled, it will be predominantly in the lungs. If injected intramuscularly, it will be mainly in the muscles around the injection site; injected intravenously, it will all be in the blood and there will be very little or nothing in the stomach and liver. If it was swallowed, then it will be mainly in the stomach, intestines and liver. Robert has noted that some jurisdictions don't collect stomach contents as standard the way they do blood samples, which he regards as unfortunate, as they 'can be incredibly useful.' In the US, too, post-mortem samples can differ widely across jurisdictions and according to the experience and discretion of the medical examiner or coroner.

Toxicology is sometimes about more than identifying alien substances in the body. It can even help reconstruct the circumstances around suspicious deaths. The moral stakes are high when somebody may have been unlawfully killed by any employee of a public institution; higher still when their job is to care for the sick and fragile.

Sister Jessie McTavish, a 33-year-old nurse, worked on a geriatric ward at Ruchill Hospital in Glasgow. On 12 May 1973 she watched an episode of the American TV series *A Man Called Ironside,* in which the relatives of elderly patients paid a nurse to murder them by lethal injection. The next day she discussed the programme with some of her colleagues, one of whom mentioned that insulin poisoning doesn't leave a trace. Three weeks after the episode, patients in Jessie's ward starting dying unexpectedly; five passed away in June alone.

On 1 July, a sixth patient, 80-year-old Elizabeth Lyon died unexpectedly. Alarm bells began to ring with the doctor who certified her death. He spoke to patients on Jessie's ward, one of whom was terrified of her. Jessie had given her an injection which had made her 'feel awful'; when asked, the nurse said that the syringe contained sterilised water, a placebo. Other members of staff revealed that Jessie was in the habit of giving patients injections without recording them in the patients' notes. Witnesses had heard her tell a visitor she was known at the mortuary as 'Sister Burke and Hare,' after the serial murderers who supplied corpses to medical schools, because of all the recent deaths on her ward.

McTavish was suspended, and charged with injecting drugs not medically prescribed into three further patients, one of whom had died. At the time the technology for measuring insulin in the body was not well developed. Nevertheless the toxicologist established that tissue from both of Elizabeth Lyon's arms showed needle marks and contained excessive insulin.

McTavish went on trial in June 1974, and was convicted of the murder of Elizabeth Lyon and of assaulting three other patients with illegal injections. A selection of nurses and doctors gave evidence against her. One nurse recorded finding three empty insulin phials in a ward side room, even though none of the patients had been prescribed insulin at that time. Another nurse testified to Jessie saying 'they could dig up the bodies if they liked and they would not find any trace of insulin'. She was sentenced to life imprisonment.

Five months later McTavish appealed her conviction. Her lawyer argued that the original judge, Lord Robinson, had misdirected the jury by not telling it that Jessie had in fact denied responding to her murder charge in the way the police inspector who made the charge said she had. That inspector had not taped Jessie's response, but he had told the trial that she said, 'I gave half a cc of insulin soluble to Mrs Lyon only because she wanted to be put out of pain and misery, and had trouble with her bowels.' Jessie denied saying this, claiming instead that she only mentioned injections of sterilised water. She said the inspector had told her that if she admitted to injecting insulin she would only get 'a £5 fine in the Sheriff's court'. The appeal judges agreed that Lord Robinson had misdirected the jury and quashed McTavish's verdict and sentence.

McTavish's name was removed from the Register of Nurses in Scotland. A short time later she married, and in 1984 she was restored to the Professional Register of the UK Central Council for Nursing, Midwifery and Health Visiting, under her married name.

Jessie McTavish's conviction was quashed. There was never any question of that happening in the case of a notorious medical practitioner who made a habit of injecting his patients with morphine and who wrote his own death certificates.

Harold Frederick Shipman (always known as 'Fred') was born in 1946 on a council estate in Nottingham. A bright boy, he did so well at his 11-plus exams that he was awarded a scholarship to the best local boys' grammar school, High Pavement. His mother always felt that she was a cut above her neighbours, and brought Fred up to feel superior, which contributed to his isolation from his peers. He was devoted to his mother, and was devastated when lung cancer slowly and painfully wrenched her from him. The doctor came in the afternoons to inject morphine to alleviate her pain, and Fred was usually there, watching his mother sink into a peaceful state. She died in 1963, when Fred was seventeen.

In his first year at Leeds University medical school, in 1965, he met Primrose Oxtoby, a 16-year-old window dresser, whom he married three months before their daughter was due. While still a student, a new husband and a first-time father, Shipman became addicted to pethidine, a painkiller used mainly in childbirth. As part of their training, the medical students were encouraged to experiment with different drugs in groups of four, two taking the drug and the other two monitoring its effects – and this is probably how he became hooked.

Shipman forged prescriptions for pethidine for years, until eventually his veins collapsed. After psychiatric treatment for his addiction, he gave it up in 1975. Outwardly he seemed a normal middle-class family man, with four children and a devoted wife. His patients thought him a good doctor and, whilst a few colleagues found him arrogant and aloof, he was generally liked in the communities in which he worked, first in Todmorden, Yorkshire from 1974, and later, from 1977, in Hyde, Lancashire.

But the truth about Shipman was that he was the diametric opposite of a good family doctor. For twenty-five years, he murdered patients at a rate of about one a month. Typically, he would visit elderly women living alone in their homes, inject them with a lethal dose of morphine, leave them sitting in a chair or on a sofa, fully clothed, and turn the fire on high. The next day, he would return, pronounce them dead and give an estimate of death considerably later than his previous visit. He managed to do this because the heat of the room kept the bodies warm, distorting body-cooling evidence of time of death. He would declare the patient dead of heart failure or old age, no post-mortem necessary because he'd recently been in attendance on them.

By 1998, some members of the Hyde community were beginning to grow suspicious. A local taxi driver, who often ferried old ladies around, noticed that they seemed to die shortly after seeing Shipman. Linda Reynolds, a nearby GP, noticed that his patients were dying three times more frequently than hers. Shipman sensed that he was being watched. He made sure that his next few victims were Roman Catholic women, who were certain to be buried rather than cremated, as, before a body could be cremated, two doctors had to examine it to make sure there was nothing suspicious that might necessitate an autopsy.

His final victim was Kathleen Grundy, 81, a former mayor of Hyde. She was, said Shipman himself, 'fit as a fiddle' when he went round to her house on 24 June to take blood for a test. The next day when she failed to turn up to help at a luncheon club for the elderly, two friends found her lying on her living room sofa, dead. They called the police, who informed Shipman. He went round to her house, gave her a quick examination and signed the certificate, writing 'old age' under cause of death. He falsified her medical notes, too, adding observations suggesting that she was abusing codeine, a cough medicine which breaks down into morphine after death, because he knew that toxicology tests were likely to find morphine.

She was buried, according to her wishes. But then a will was produced that left her entire £380,000 estate to Harold Frederick Shipman. 'I give all my estate, money and house to my doctor,' the will read. 'My family are not in need and I want to reward him for all the care he has given to me and the people of Hyde.' When Kathleen's daughter saw the will she was shocked, finding it 'inconceivable' that her mother had written it. She alerted police, who ordered the body to be exhumed and a post-mortem

KATHLEEN GRUNDY
LOUGHRIGG COTTAGE
79 JOEL LANE
HYDE
CHESHIRE
SK14 5J

RECEIVED 24 JUN 1998

22.6.98

Dear Sir,

I enclose a copy of my will. I think it is clear in intent. I wish Dr. shipman to benefit by having my estate but if he dies or cannot accept it ,then the estate goes to my daughter.

I would like you to be the executor of the will, I intend to make an appointment to discuss this and my will in the near future.

Yours sincerely

K. Grundy.

Serial killer Harold Shipman, and (inset) a letter accompanying the forged will of his final victim, Kathleen Grundy. The letter was later matched to a typewriter in Shipman's surgery

performed. Meanwhile, investigators discovered one of Shipman's fingerprints on the will and linked it to a battered old Brother manual typewriter that he kept in his surgery.

Kathleen was exhumed on 1 August, six weeks after her funeral. Forensic pathologist Dr John Rutherford carried out an autopsy, but could find no obvious cause of death. He sent the muscle from her left thigh and a liver sample to forensic toxicologist Julie Evans. The thigh muscle is the most stable tissue in the body, making it a good place to find traces of poison. Julie Evans tested the thigh and liver using mass spectrometry, which produces a graph showing the levels of different chemicals in the sample. On 2 September she reported that Kathleen Grundy had died from a fatal dose of morphine.

Morphine is medical heroin, a strong and highly addictive painkiller usually only prescribed for patients in the final stages of a terminal illness. Shipman had been stockpiling it through false prescriptions and by secretly taking it from cancer patients after they had died. The drug acts on the central nervous system, alleviating pain and inducing peacefulness. If it is injected into a vein, breathing slows instantly, followed by loss of consciousness, and death. As murder goes, it's quick and painless. But it's still the violent sundering of a person from their life.

Understanding that morphine remains in bodies for a long time after death, the coroner ordered a further eleven of Shipman's patients to be exhumed. All contained lethal quantities of morphine. Shipman was arrested and put on trial on 4 October 1999, charged with fifteen murders and the forging of Kathleen Grundy's will. He was sentenced to life imprisonment, but in 2004 he made a noose from a bed sheet and hanged himself from the bars of his cell window.

The Shipman Inquiry, chaired by High Court Judge Dame Janet Smith, examined all the deaths among Shipman's patients during his entire career – 887 in all. Smith's final report in 2005 estimated that Shipman had murdered 210 of his patients, with a possible further 45, making him the most prolific killer ever convicted. Though the majority of his victims were elderly, there was 'quite serious suspicion' that he had killed one patient aged four. That he was not caught sooner provoked anger among the public, and soul-searching among the medical and forensic community.

How did Fred Shipman become such a calculating monster? And why? Until Kathleen Grundy, Shipman had never killed for financial gain. Probably, we will never know: he took all his conscious motivations with him to the grave, lying about his murderous ways until the end.

Perhaps he was influenced by Dr John Bodkin Adams, who was charged in 1957 with using morphine to murder 160 of his wealthy patients in Eastbourne, Sussex. (Although he was acquitted, in recent years opinion has tended towards the conclusion that he was probably guilty.) However, psychologists propose that the afternoons Shipman spent observing the calming effects of morphine on his mother as a factor. Robert Forrest, who has written several papers on healthcare murderers, makes the point that doctors are a cross-section of society and have 'no special charisma of virtue'. Some common reasons to work in health care include intellectual curiosity, altruism, the promise of social status and financial security. Robert has estimated that about one in a million enters based on darker feelings, 'with a more overtly disturbed psychology, where the perpetrator is thrill-seeking or even actively psychotic'. For someone like Shipman, 'being able to manipulate and control patients up to the point of killing them is interesting'. Given the arrogance he showed to police officers who interviewed him in the days after his arrest, given his belief that he had the right to privately murder whoever he wanted, it seems that Shipman enjoyed exercising the power of life and death, and thought he could play God for ever.

Thankfully the majority of would-be murderers do not have access to the same armoury of drugs as medical professionals. They are also unlikely these days to turn to metallic poisons like arsenic because they are easy for modern toxicologists to identify. Their poisons of choice are plant-based, and they sometimes administer them in barely credible ways. As a writer, I was inspired by the Poison Garden at Alnwick Castle to create a serial killer whose fascination with vegetable poisons proved fatal for several victims.

But stranger than anything I could have invented is the case of Georgi Markov. On 7 September 1978 Markov was standing at a bus stop on Waterloo Bridge in London when he felt a sharp pain in the back of his right thigh. Markov was a Bulgarian dissident writer who had defected to the West in 1969. He was waiting for a bus to take him to his job at the BBC World Service, where he broadcast shows satirising the Communist regime in Bulgaria. Markov flicked his head round and noticed a man close by pick up an umbrella from the ground, hail a taxi and disappear. He felt like he'd been stung by a wasp or a bee. When he reached his office, he noticed a small red pimple on his leg. Later that evening, his leg became inflamed and he developed a fever. Next morning an ambulance took him to hospital. The doctors X-rayed his leg but found nothing. Despite large doses of antibiotics, Markov died four days later.

The coroner suspected poisoning and ordered an autopsy. The pathologist, Rufus Crompton, found nearly all of Markov's organs to be damaged, and confirmed that he had died of acute blood poisoning. He also found a tiny pellet, the size of a pin head, just underneath the skin of Markov's thigh, with two tiny holes drilled into it.

Crompton sent the pellet and the tissue surrounding it to toxicologist David Gall, who ran tests but could not identify the poison. But based on the sequence of Markov's symptoms, he thought the bullet might contain ricin, a substance extracted from the seeds of the castor oil plant – 500 times more powerful than cyanide. Harking back to Mathieu Orfila and his tests on dogs, Crompton decided to inject a pig with ricin. 'It had exactly the same symptoms,' he observed. 'It died in the same way; samples of its blood showed the same high white cell count that no other poison produces.'

If ricin is swallowed, its symptoms are nasty but not fatal. But if it is injected or inhaled or absorbed through the mucus membranes, a dose the size of a few grains of salt will kill an adult man. Ricin inhibits the protein synthesis of cells, causing cell death, and damage to the major organs. There is a delay of a few hours before the appearance of symptoms, which include high fever, seizures, severe diarrhoea, chest pains, breathing difficulties and oedema. Death ensues within three to five days; there is no antidote. Poisoners have favoured it over the years because, like arsenic, its symptoms mimic natural causes.

In the Markov case Crompton supposed that someone had drilled into the pellet, inserted a few grains of ricin and sealed it back up with a sugary coating designed to melt at 37 degrees, the temperature of the human body. To fire the pellet, the assassin must have used a mechanism that worked like an air rifle and looked like an umbrella. There had been a similar attempt, using the same kind of pellet, on the life of a Bulgarian defector in Paris ten days before Markov was hit, but the victim had survived because the coating on the pellet had only partially melted.

Because there had been two previous assassination attempts on Markov, the police suspected that his murder was orchestrated by the Bulgarian secret police, probably aided by the Russian KGB. In 1990, Oleg Gordievski, a double agent, claimed that the KGB had supplied the poison and made the umbrella gun. The Soviet Union fell in 1991 and in the following year the former head of the Bulgarian intelligence service destroyed an archive of ten volumes which gave details of assassinations ordered by the regime. Exactly who killed Markov will probably never be known.

Ordinary civilians tend to administer their plant poisons of choice in less elaborate ways. In 2008 in Feltham, west London, 45-year-old Lakhvir Singh, a mother of three, had been dumped by Lakhvinder Cheema, her lover of the past sixteen years. Cheema, known to his friends as 'Lucky', had started seeing a woman half her age. Lakhvir was heart-broken. Then Lucky announced he would marry his new girlfriend on Valentine's Day. Lakhvir decided that she would rather kill him than suffer the perpetual anguish of knowing he was with another woman. So she went to Bengal for a month, to the foothills of the Himalayas, and came back with poison extracted from the beautifully flowering monkshood plant, also known as Indian aconite,

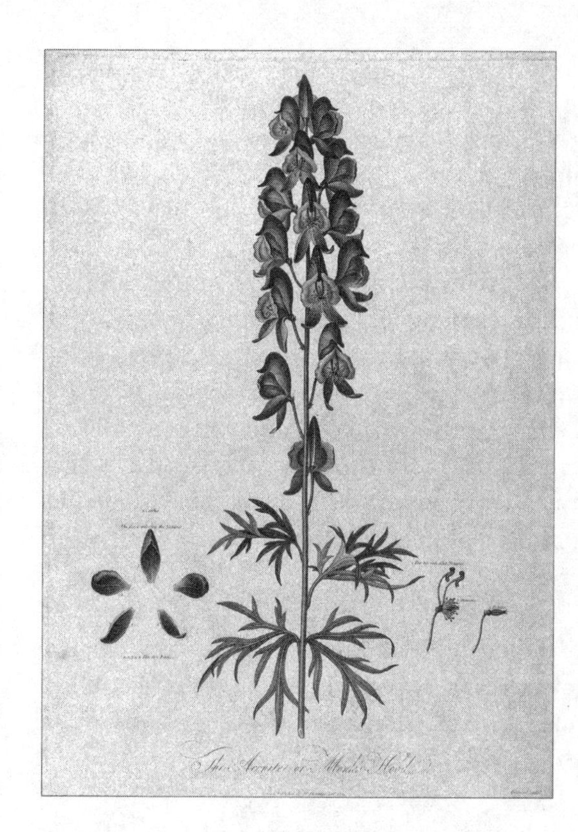

Aconite, also known as monkshood and wolfsbane. Symptoms of aconite poisoning include nausea, vomiting, burning and tingling in the limbs, and difficulty in breathing. If untreated, death can occur within two to six hours

thought to be the deadliest plant in the world. (Incidentally, in J. K. Rowling's *Harry Potter and the Half-Blood Prince,* Professor Snape uses the British version, wolfsbane, to stop Remus Lupin turning into a werewolf.)

On 26 January 2009, two weeks before the planned wedding, Lakhvir Singh let herself in to Lucky's house in Feltham, took a container of leftover curry from the fridge and laced it with aconite. Lucky and his fiancée ate the curry for dinner the next day. He enjoyed it so much that he had a second helping. Shortly afterwards, both of them started vomiting. His fiancée recalled what happened next: 'Lucky said to me, "I am not feeling very well. My face has become numb and, when I touch it, I cannot feel it."' Next he started losing the use of his arms and legs. He managed to call 999 and told the operator that he thought his ex-girlfriend had poisoned his food. The couple were rushed to hospital, where Lucky died.

Aconite stops the heart and other internal organs from working. After the severe vomiting, the victim feels like they have ants crawling over their body, then they lose sensation in their limbs, their breathing becomes slower and slower, and their heartbeat weakens, disturbing the rhythm of the heart. However, the mind remains clear throughout. Lucky's fiancée was put in a drug-induced coma for two days while toxicologist Denise Stanworth tried to trace the poison. Robert Forrest explains: 'Fortunately Denise had enough post-mortem material to go on. It was only when she started looking for exotic vegetable poisons that she found the aconite.' The woman was given a dose of digitalis, which calmed her abnormal heart rhythm, and she went on to make a full recovery.

When the police searched Lakhvir Singh's flat, they found two packets of brown powder containing aconite in her coat and handbag. She claimed it was medication for a rash on her neck, but was found guilty of murder and sentenced to twenty-three years in prison.

Sometimes toxicologists are confronted with poisons before they have entered the body. The fire scene investigator Niamh Nic Daéid, whom we met in chapter 2, is also an analytical chemist who specialises in fire, explosives and drugs. When Niamh wants to know if cocaine is present in something, she first uses a simple yes-or-no colour test. 'We stick it in a little tube, shake some reagent on it and if it goes blue then it's cocaine.' Next she employs more sophisticated techniques, such as gas chromatography, to find out how concentrated the drug is.

When a researcher from Thailand came to visit, she explained to Niamh that poorer countries can't afford this second round of tests. Niamh realised that people were being arrested on the basis of the colour test alone, regardless of the concentration of the drug. So Niamh and her team devised a cheaper solution. 'You take a photograph of the colour using a smartphone and then, once you've calibrated the camera, you use the colour to give a tentative percentage of drug that's present in the sample. Because you've taken it with a smartphone, the image has GPS co-ordinates attached, and you can beam it up. We're now working with the United Nations to create a living global map of point-of-seizure drug samples. A lot of the frontline forensic science work that makes a difference in a global sense doesn't have to use complicated technology. It can be actually a very simple solution to a problem.' A Pantone test for cocaine concentration may not have been what Mathieu Orfila had in mind two centuries ago, but I can't help feeling that he'd have enjoyed its elegance.

SIX

FINGERPRINTING

'And he gave unto Moses two tables of testimony, tables of stone, written with the finger of God'

Exodus 31:18

The governing principle of forensic science, as laid down by Edmond Locard at the beginning of the last century, is that 'every contact leaves a trace'. But unless we know how to analyse, categorise and understand those traces, they're not much use when it comes to catching criminals. As scientists have made new discoveries, so the art of detection has advanced. And the technique of identification from fingerprints was a headline-grabbing trailblazer in terms of bringing criminals to justice

Forensic science did not begin with fingerprinting, but it caught the public imagination in a way that no other development had. And because it was so easy to comprehend, the courts also took to it readily. In the early 1900s the law-abiding citizenry fell in love with the idea that a silent burglar who touched what wasn't his could be identified just as silently; that the murderer who took a life with a blunt instrument could swing from the gallows thanks to the pattern on the tip of his pinkie; that a moment's carelessness would lead inevitably to conviction thanks to the unique arrangement of a clutch of ridges and loops.

One of the first Europeans to grasp the idea of the individuality of fingerprints was a young man named William Herschel. In 1853 he set sail from England to work for the East India Company, which effectively ruled large parts of India. Four years later a dispute over the type of grease used in gun cartridges led to a group of the company's Indian soldiers mutinying against their British commanders. The subsequent rebellion – known as the Indian Mutiny – spread across the country, leading to widespread violence met with vicious reprisals from the British forces. When the dust settled, the East India Company was forced to turn over its territories to the British Crown and many of the company's employees were transferred to the Indian Civil Service. Herschel was put in charge of a rural region in Bengal.

The brutality of the rebellion had left feelings running high and many Indian citizens were determined to make life as difficult as possible for their British overlords. They stopped turning up for work, paying taxes and cultivating British farms.

Herschel, an ambitious 25-year-old, was determined not to let civil unrest stand in the way of his making a mark. One of the first decisions he took in his new role was to build a road. He drew up a contract with Konai, a local man, to supply equipment for the project. Then he did something very odd.

'I dabbed Konai's palm and fingers over with home-made oil-ink used for my official seal, and pressed the whole hand on the back of the contract, and we studied it together, with a good deal of chaff about palmistry, comparing his palm with mine on another impression.' When Herschel printed Konai's hand, he wasn't thinking of it in terms of identification, but as a kind of insurance – 'to frighten him out of all thought of repudiating his signature'.

Although rare, and by 1861 illegal, Herschel may have come up with the idea for the handprint from the Hindu practice of suttee, whereby a widow was burned alive on the funeral pyre of her dead husband. As she passed through the 'Suttee Gate' on her way to death, the doomed woman would dip her hand in red dye and place it on the Gate. The stonework around the print would then be carved away to make it stand out in bas-relief.

Twenty years later, Herschel was appointed magistrate at Hooghly, near Calcutta, where he was responsible for the courts, the prison and pensions. We think of benefit fraud as something modern, but Herschel was aware of it 140 years ago. He set up a system for taking the fingerprints of pensioners so that when they died other people couldn't fraudulently collect their pensions. He also took the prints of people when they were given jail sentences, to stop convicted criminals paying others to serve their prison sentences for them.

The idea of being able formally and categorically to identify criminals was gathering traction in various jurisdictions. Around the same time that Herschel was developing his system, a police clerk in Paris called Alphonse Bertillon was buckling under the volume of incoming prisoners. He decided to try to identify them systematically using anthropometry, the science of measuring humans. He chose eleven body measurements, including width of head and distances from elbow to end of middle finger. Bertillon put the odds of two people sharing all eleven of his measurements at one in 286 million. He recorded the individual measurements on file cards, and in the middle of the cards he stuck two photographs – full face and profile – and thus the mug shot was born.

Meanwhile, near Tokyo, a Scottish medical missionary started experimenting with fingerprints. Henry Faulds had noticed that ancient potters marked their pots with finger and thumb marks. He also discovered that subtle prints could be made visible when dusted with powder, and he used the technique to exonerate a man accused of burglary. When Faulds showed the real thief the similarities between his print and the one on a windowpane in the burgled house, he broke down and confessed. As a result of his observations, Faulds devised a method of fingerprint classification based on prints from all ten fingers. He tried to persuade Scotland Yard to set up a fingerprint department using his system, but he was rejected.

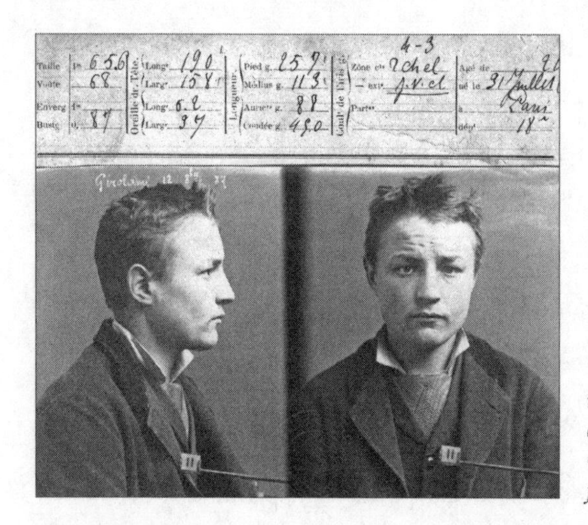

The Bertillonage record of twenty-one-year-old George Girolami, arrested for fraud

Undaunted, Henry Faulds wrote to Charles Darwin detailing his fingerprinting method. Darwin was intrigued by the idea but felt that it was work more suited to a younger man, so he passed it on to his cousin, Francis Galton. Galton spent ten years studying fingerprints and wrote the first book on the subject, *Finger Prints* (1892), in which he distinguished eight basic fingerprint patterns of arches, loops and whorls. He also demonstrated that every human finger fits into one of these categories in its own unique way.

After reading an article by Galton, Croatian-born police official Juan Vucetich began collecting the fingerprints of arrested men in Buenos Aires, Argentina. He devised his own ten-finger system of classification which he called 'dactyloscopy' – and which many Spanish-speaking countries still use today. As well as being used in criminal cases, it was quickly adopted by the Argentinian government as a form of verification on internal identity passes.

Vucetich's system soon faced a demanding and disturbing test case. On 29 June 1892, in a village near Buenos Aires, 4-year-old Teresa Rojas and her 6-year-old brother Ponciano were found brutally murdered in their home. Their mother, Francisca, was alive but her throat had been cut.

Francisca told the police that her neighbour, Pedro Velázquez, had stormed into her home, murdered her children and tried to slit her throat. The police tortured Velázquez for a week, but he stuck to his alibi: he had been out with a group of friends at the time of the murders.

Frustrated at the lack of a confession, Inspector Commissioner Alvarez returned to the house. This time he noticed a brown patch on a doorframe, which he thought might be a bloody fingerprint. He removed the bloodstained section of wood from the frame and took it, along with fingerprints taken from Velázquez, to Juan Vucetich, who had just opened a fingerprint bureau in Buenos Aires.

Vucetich confidently declared that the prints did not match those on the doorpost. Then he took the prints of Francisca Rojas. They were identical. Faced with the damning bloody fingerprint evidence, the mother confessed to murdering her two children, cutting her own throat for effect and accusing an innocent man. She had wanted to improve her chances of marrying her boyfriend, who didn't like children. Instead she became the first person to be convicted of a crime based on fingerprint evidence. She was sentenced to life imprisonment.

Following the Rojas case, Argentina abandoned Bertillon's anthropometric system and began organising its criminal records on fingerprints alone. Soon, other countries began to follow suit. The following year, Edward Henry, the chief of police in Bengal, added thumbprints to his anthropometric criminal records. Although fingerprinting had been used officially in civil matters in Bengal since William Herschel had introduced the system forty years previously, the police had never taken advantage of it. Working with an Indian police officer, Azizul Haque, Henry improved on Galton's system to produce one which allowed an investigator to use the physical characteristics of a fingerprint to create a unique reference number. These numbers were then used to file the fingerprints in one of 1024 pigeonholes in the police headquarters; when a new fingerprint was taken, its characteristics were coded, and the appropriate pigeonhole checked to see if it had been filed before. In 1897, the 'Henry Classification System' was adopted throughout British India.

In 1901 Henry was recalled to London to head the Criminal Investigation Department (CID) at Scotland Yard. He immediately set up the Fingerprint Bureau to record the identity of criminals in the belief that it would foil repeat offenders. Before there was a reliable system of recording their identities, it was common practice among career criminals to dodge a harsher sentence by making up an alias and pretending to be a first-time offender rather than a recidivist. In its first year alone the bureau cracked the pseudonyms of 632 habitual criminals.

As is often the case with new developments, it takes a sensational case to establish a new forensic technique in the public consciousness. For fingerprints, that leap into the forensic limelight came four years later, in 1905. On a Monday morning at the end of March, William Jones turned up for work at Chapman's Oil and Colour Shop in Deptford High Street, London. The 16-year-old was surprised that, although it was half-past eight, the paint shop was locked and shuttered. The manager and his wife lived above the shop and they usually opened up for their early customers at half-past seven. William was concerned that they might be ill; at seventy-one and sixty-five, it seemed plausible. When knocking produced no response, William barged the door with his shoulder. It held fast. He stood on tiptoe and squinted through a gap in the shutters. At the back of the shop, by the fireplace, he could make out a lounge chair tipped on its side.

Anxious now, William ran to fetch a friend. Together they hurried back and broke the door down. William discovered his boss, Thomas Farrow, lying underneath the overturned chair, his bald head smashed open, blood seeping into the ashes of the fireplace. During the subsequent autopsy, the pathologist expressed the view that Thomas had been hit six times around his head and face, probably with a crowbar.

Sergeant Albert Atkinson was the first policeman on the scene and it was he who found Ann Farrow in bed upstairs, grievously battered and unconscious, but still clinging on to life. Atkinson noticed a moneybox lying open and empty on the floor by the Farrows' bed. William explained that Mr Farrow normally took the box to the bank on Mondays to deposit the shop's weekly takings of around £10.

Melville Macnaghten, Edward Henry's successor as head of the CID, took charge of the case. On his first day at Scotland Yard, in 1889, Macnaghten's new boss had briefed him on the unsolved Jack the Ripper murders of the previous year. For the rest of his career Macnaghten kept photographs on his desk of Jack the Ripper's mutilated victims to remind him to try harder. Still, like any experienced detective, he had unsolved murders of his own. Just three days after he joined the CID, Macnaghten had found himself picking up pieces of a dismembered woman from along the riverbank. Her murderer was never found and the case became known as the 'Thames Mystery.'

Macnaghten was determined to solve Thomas Farrow's brutal murder. The crime had shocked local residents. Deptford was a polluted and crowded district. Disease and crime were facts of life. But cold-blooded murder was rare.

Because the elderly couple had been found in their nightclothes, and because the pathologist estimated Thomas Farrow had died not long before William discovered his body, the police believed that Thomas had been tricked into opening his front door early in the morning. They speculated that the assailant had attacked Thomas immediately, then run upstairs to find his moneybox. Detectives found a pool of blood at the top of the stairs, so they deduced that Thomas had somehow managed to chase his attacker upstairs, where his wife lay unprotected. They concluded that the assailant had callously finished him off, silenced his wife with complete ruthlessness, taken the money and run.

Macnaghten examined the moneybox carefully and noticed a greasy finger mark on the underside of its inner tray. He picked the box up with his handkerchief, wrapped it in paper and took it

A CID assistant checks a new set of prints against Scotland Yard's fingerprinting records in 1946

to the Fingerprint Bureau. Macnaghten knew that he was risking public ridicule because, although fingerprint evidence had nailed a burglar, Harry Jackson, in 1902, it still had the taint of palmistry about it. Not everyone had been convinced by the efficacy of fingerprinting at Jackson's trial. After hearing the guilty verdict, someone signing off as 'A Disgusted Magistrate' wrote a letter to *The Times*: 'Scotland Yard, once known as the world's finest police organisation, will be the laughing stock of Europe if it insists on trying to trace criminals by odd ridges on their skins.'

Charles Collins, head of the Fingerprint Bureau, looked at the tray with a magnifying glass and, from the size of the print and the slope of its ridges, recognised it as the right thumb of a sweating hand. He was also pleased to see significant differences when he compared it to the prints of Sergeant Atkins and the Farrows, which he had also taken. Those differences would strengthen the case against a suspect whose right thumb matched the print.

Although the bureau was only four years old, it had already amassed around 90,000 finger and thumbprints, stored in its enormous wooden cabinet of pigeonholes. Collins looked in the appropriate pigeonhole, but found no match.

The next blow to the investigation came five days later, when Ann Farrow died from her injuries. Macnaghten had hoped she would return to consciousness and describe her attacker.

Then there was one of those breaks that sometimes come detectives' way thanks to the media. Having read about the murders in the press, a milkman came forward, saying that he had spotted two men leaving the Oil and Colour Shop at 7.15 a.m., and had shouted out to them that they had left the front door ajar. One of them had turned round and said, 'Oh! It don't matter' before they had both walked off. The milkman described their appearance. One had a dark moustache and was wearing a blue suit and a bowler hat; the other was wearing a brown suit and a cap.

Then another eyewitness, a painter, came forward to explain why young William Jones had found the front door locked. The painter had glimpsed an old man with blood on his face shutting the door, at 7.30 a.m. Macnaghten reasoned that Thomas Farrow had somehow survived a second battering, this time at the top of the stairs, and, in a state of deliriousness, had staggered downstairs and closed the door, before moving to the back of the shop and finally succumbing to his wounds.

A third eyewitness came forward: a woman who had seen two men matching the milkman's description running down Deptford High Street at 7.20 a.m. Even better for the police, she recognised one of them. The man in the brown suit, she said, was 22-year-old Alfred Stratton. The descriptions of his companion matched that of Alfred's 20-year-old brother, Albert. When the police went to question Alfred's girlfriend, she admitted that the day before the murders Alfred hadn't had enough money to buy food; but that the day after, he had come back with bread, bacon, wood and coal. This was enough for Macnaghten. The Stratton brothers were arrested a week after Thomas Farrow's murder.

But the ill luck that had dogged the investigation continued. Neither the milkman nor his assistant was able to pick the Strattons out in a line-up. The brothers were filled with bravado; they joked that, when Charles Collins took their fingerprints, it tickled them.

Collins had the last laugh, however. When he examined the fingerprints, he found the mark on the moneybox matched Alfred Stratton's right thumb.

Still the prosecution knew they had their work cut out. Could a one-inch sweat mark persuade the jury? So much hung on this case: the conviction of a pair of cold-blooded killers; the restoration of the reputation of Scotland Yard, so damaged by Jack the Ripper's murders; and the acceptance of fingerprints as key evidence. Both Macnaghten and the Metropolitan Police Chief, Edward Henry, understood how much was at stake.

Ironically, Henry Faulds, back from Japan, was ready to testify for the defence. He had his own axes to grind. First of all, Scotland Yard had rejected his calls to set up a fingerprint bureau. Then it had opened one up based on the Henry system, and refused to acknowledge Faulds' part in the development of fingerprinting. Faulds was intent on claiming that not enough research had been done to prove that a print from one finger could identify an individual beyond doubt.

Charles Collins took the stand, several blown-up photographs under his arm. He showed the jury the smudged print taken from the moneybox, then the perfectly crisp prints of the Farrows and Sergeant Atkinson. The jury didn't need much explanation to see that the prints were different from each other. Then Collins produced Alfred Stratton's thumbprint. The similarity was immediately obvious. Collins pointed out eleven separate points of similarity. The jurors were mesmerised.

When the defence cross-examined Collins he argued cogently that no two prints from the same finger are ever exactly the same, because of differences in the pressure and angle of contact. That was just as well, for, when the first fingerprint expert for the defence, John Garson, took to the stand, he proceeded to discredit

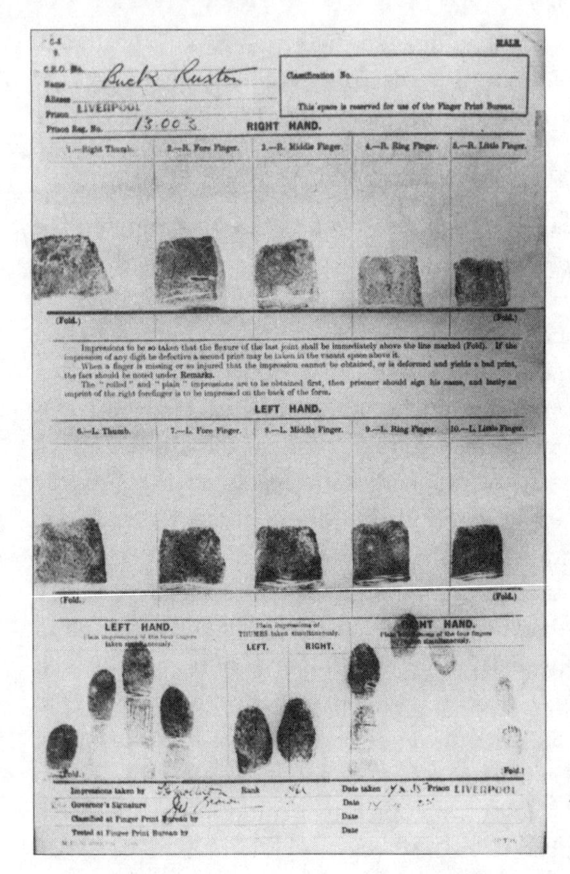

MALE.

C.R.O. No.
Name Buck Ruxton
Aliases
Prison LIVERPOOL
Prison Reg. No. 13.003

Classification No.
This space is reserved for use of the Finger Print Bureau.

RIGHT HAND.

1.—Right Thumb.	2.—R. Fore Finger.	3.—R. Middle Finger.	4.—R. Ring Finger.	5.—R. Little Finger.

(Fold.) (Fold.)

Impressions to be so taken that the flexure of the last joint shall be immediately above the line marked (Fold). If the impression of any digit be defective a second print may be taken in the vacant space above it.

When a finger is missing or so injured that the impression cannot be obtained, or is deformed and yields a bad print, the fact should be noted under **Remarks**.

The "rolled" and "plain" impressions are to be obtained first, then prisoner should sign his name, and lastly an imprint of the right forefinger is to be impressed on the back of the form.

LEFT HAND.

6.—L. Thumb.	7.—L. Fore Finger.	8.—L. Middle Finger.	9.—L. Ring Finger.	10.—L. Little Finger.

(Fold.) (Fold.)

LEFT HAND. Plain impressions of the four fingers taken simultaneously.

Plain impressions of THUMBS taken simultaneously. LEFT. RIGHT.

RIGHT HAND. Plain impressions of the four fingers taken simultaneously.

(Fold.) (Fold.)

Impressions taken by
Rank
Date taken
Prison LIVERPOOL
Governor's Signature
Date
Classified at Finger Print Bureau by
Date
Tested at Finger Print Bureau by
Date

Buck Ruxton's fingerprints, taken in Liverpool Prison, 1936

Collins' eleven points of similarity. The distances between the points were slightly shorter in some and longer in others, he said. And the lines curved a tad differently as they reached the points.

The eminent counsel for the prosecution Richard Muir started his cross-examination of Garson by placing two letters down in front of him. They had both been written by Garson on the same date. Garson had written one to the Stratton lawyers, offering to testify for them. The other was to the Director of Public Prosecutions, making the same offer. Muir's implication was that Garson was a hired gun, willing to sell his services to the highest bidder. To this charge, Garson replied, 'I am an independent witness.' To which the judge added sternly, 'An absolutely untrustworthy one.' Garson got down from the stand with his credibility in tatters.

Henry Faulds was due to give evidence next. He was ready to deliver his devastating view that, after comparing many thousands of fingerprints, he could not definitively say that a single fingerprint could only belong to one person on earth. But the defence panicked lest Muir would undermine him as successfully as he had Garson. Faulds never got to speak.

After a two-hour deliberation, the jury came back with a guilty verdict. On 23 May 1905, nineteen days after their trial had begun, the Stratton brothers both went to the gallows. The British judicial system had stepped into a whole new realm of scientific evidence.

By 1905, fingerprint bureaus had been established in India, the UK, Hungary, Austria, Germany, Switzerland, Denmark, Spain, Argentina, the US and Canada, but their evidence had so far only been used as proof of guilt in Buenos Aires and London. The Stratton case demonstrated how powerful this evidence could be. In 1906, the year after that keynote trial, four other British men were prosecuted based on crime scene prints. In the same year, the New York City Police Department (NYPD) introduced fingerprinting to police departments across the US.

Edward Henry's system of classifying and finding matching prints remained essentially unchanged until computers led to the automatic fingerprint identification system in the 1980s. And so has the work of the fingerprint examiner.

First of all an examiner needs to understand what they are looking at. The pads of our fingers are home to intricate patterns of ridges and valleys. If we coat one of those pads with ink and press it on to a piece of paper, the resulting ridge pattern is instantly recognisable as the iconic image of a fingerprint. Our fingerprints are part of us from before birth; they first appear in the tenth week of pregnancy, when the foetus measures only 8 cm. As one of the three layers of tissue that make up the foetus's skin – the basal layer – starts to grow at faster rate than the other two, ridges form to relieve the resulting stresses, 'like the buckling of land masses under compression'. If your finger pads were flat, the pressure on the skin would be equal and the ridges would be parallel. But because finger pads slope, ridges form along lines of equal stress, most usually in concentric circles. Ridge patterns also appear across the palms of our hands and on the bottoms of our feet. Other primates have them, too, and evolutionary biologists believe there are good reasons for them. They help our skin to stretch and deform, protecting it from damage; they create valleys down which sweat can escape, reducing slipperiness when we hold things; and they give us more contact (and hence grip) with rough surfaces like tree bark.

When we touch a surface with a finger, the ridges leave their unique pattern on it. Even the prints of identical twins differ. In all the years that fingerprinting has been practised, no one has yet found two identical clear and complete prints belonging to two different fingers.

To identify people from the marks they leave behind is simple enough in a family setting. These small muddy prints belong to the toddler who forgot to take his shoes off. Those even smaller ones are the dog's. But deductions like these are made from a small number of possible culprits, and the prints in question are patent – visible to the naked eye. Invisible, latent prints are much trickier. Substances like sweat, mud, blood and dust can produce both patent and latent prints. The more absorbent or irregular the surface, the more difficult it is for a CSI to get a good print. Though it was once impossible to take prints from plastic bags or human skin, the technologies have improved and there are now ways to do it.

CSIs use various methods for recovering prints in a logical sequence, starting with the least destructive. In Britain, the order of actions is set out in the Home Office's *Manual of Fingerprint Development*. First the CSI examines surfaces for patent prints, like the bloody one on the doorframe at the Rojas' house; if necessary, she photographs them. Then she shines lasers and ultraviolet light on to surfaces to illuminate latent marks, and make them photographable. If special lighting doesn't work, she delicately brushes a dark powder over the mark, takes a photo, and then presses an adhesive strip over it. The strip is peeled off and pressed on to a white card. This is Henry Faulds' classic way of recovering fingerprints from a crime scene, and it is still the most commonly used method today. If a print remains stubbornly invisible, as it might well do on more porous surfaces, the CSI can use a range of chemicals which react with the salt and amino acids in human sweat, to make it visible.

The photographs and latents are then sent to a latent print examiner, who decides whether they contain enough ridge detail to make identification possible. If the print isn't too smudged or incomplete, the officer compares it to fingerprints of neutrals – that is, people who had every right to be there and who are not suspects – including victims and police officers, before looking at prints from potential suspects. This is inevitably a subjective process. If the officer judges that none matches, she scans the print and encodes it into a geometric pattern. Then she runs an automatic search using a national database, such as the UK's IDENT1, which holds the prints of around 8 million people, or its US equivalent IAFIS, managed by the FBI, which holds all fingerprints collected by thousands of law enforcement agencies, over 100 million prints.

IDENT1 and IAFIS are the modern equivalent of Edward Henry's pigeonholes. Both databases use a slightly modified version of Henry's classification and identification system. A computer program asks the fingerprint a series of questions, such as, 'How many whorls have you got?' Each answer is given a numerical value – 'two whorls' is two points. The values are strung together to give the print an overall code. The database then compares this code to the millions of others it contains, and presents the officer with the closest matches, which can be anywhere from one to fifty depending on the clarity and type of the print being matched, and the search parameters.

The next step is usually for the examiner to judge if any of them are a positive match, though increasingly, 'lights-out' algorithms adopted for civil identifications, where human input and subjective judgment is minimized, are being tested in criminal law contexts. Once the examiner has found a similarity in the overall pattern of the ridges, she concentrates on tiny distinguishing points known as 'minutiae', which include points where ridges start, end and join together; where they sit independently; and where they form little bridges between two other ridges.

In 1901, when Scotland Yard established its Fingerprint Bureau, officers like Charles Collins needed to find at least twelve minutiae in agreement before they could testify to a match in an English court. In 1924 this was increased to sixteen points, which was higher than in most other countries. At the time most fingerprint experts thought that eight was enough. If an officer found between eight and fifteen points, they usually reported it to the police, because it might give them a valuable lead. But by 1953 all the UK police forces had adopted the 16-point standard.

Since the Stratton brothers case, belief in fingerprinting has gone from strength to strength amongst the civilians, judiciary and police forces of the world. And for many, including large numbers of experts, it has an aura of absolute infallibility. As Jim Fraser writes in *Forensic Science* (2010): 'In the view of most marks examiners, identification of an individual by fingerprints can be done unequivocally, that is, with 100% certainty.'

When a print is clear, the chances of an officer making a wrong call are next to nothing. But when a print is smeared, overlaid with other marks, made in blood, one officer may see points of agreement that another does not. A case in 1997 tested the subjective nature of fingerprinting to breaking point. On 6 January, the body of Marion Ross was discovered in her house in Kilmarnock, Scotland. She had been the victim of a horrific attack: her injuries included multiple stab wounds and crushed ribs, and a pair of scissors was found embedded in her throat. CSIs set to work recovering evidence, and found more than 200 latent finger marks in Marion's house, which they sent to the Scottish Criminal Records Office to be checked off against the neutrals – paramedics, doctors, police officers.

The finger mark that became the eye of a perfect storm was a left thumbprint on the bathroom doorframe. Despite it being quite badly smudged, a fingerprint officer confidently identified it as belonging to 35-year-old Detective Constable Shirley McKie, who was supposed to have remained outside the house to preserve the scene while investigators gathered evidence inside. She would have had to leave her post in order to touch the door; that would have been gross misconduct.

Officers are thoroughly trained how to treat crime scenes; CSIs always wear protective gloves so they don't damage the fragile traces left by criminals. Because of the seriousness of the case, a further three experts at the Scottish Criminal Records Office examined the thumbprint, and confirmed it to be McKie's. It looked as if the detective had indeed abandoned her post.

Meanwhile, the prime suspect for the murder had been identified as David Asbury, a 20-year-old handyman. Investigators had found his finger marks in Marion's house, and her prints on a tin box in his. Asbury explained that he had recently done a job in Marion's house, hence the marks. But detectives thought there was enough evidence to arrest him.

At Asbury's trial, McKie testified that she had not been inside Marion Ross's house at any time, so the thumbprint could not have been hers. All the other fifty-four officers who had worked the crime scene corroborated her assertion. Nevertheless she was suspended from Strathclyde Police, and eventually dismissed.

But that wasn't the end of her nightmare. Early one morning in 1998 Shirley McKie was arrested. She had to get dressed under the watchful gaze of a policewoman. The officers took her to the police station where her own father, Iain McKie, had been commanding officer. She was strip-searched and shut in a cell. She learned she was to be charged with perjury, which carries the threat of an 8-year jail sentence. Her father's long and distinguished police career had left him convinced of the integrity of fingerprint evidence. It was easier to believe that his own beloved daughter was lying than to doubt the experts. 'People have been hung on a fingerprint,' he reminded her.

In May 1999 Shirley McKie was brought to trial at the High Court of the Justiciary. Two American experts who had examined the thumbprint maintained that the print was not hers. One said the 'obvious' differences took 'only seconds' to see. On this evidence the jury found McKie not guilty of perjury. In August 2002, David Asbury's conviction for murder was also quashed by the Court of Criminal Appeal in Edinburgh – because of faulty fingerprint evidence. He had spent three and a half years in jail.

After Shirley McKie's innocence had been established, the Scottish Criminal Records Office and four police officers of the Strathclyde Police were accused of misconduct. McKie subsequently launched a damages case and in 2006 received a £750,000 settlement.

But by then she had lost the job that she loved, spent years working in a gift shop and suffered from severe depression. Iain McKie now travels the world campaigning for better expert evidence to be produced in courts of law, and warning people about the entrenched attitudes of fingerprint experts.

In 2001, the 16-point standard was scrapped in England and Wales, partly because of the McKie–Asbury fiasco, and partly because it wasn't really a standard. If fingerprint officers had found fourteen points they would sometimes search for two more, to get a 'match'. They were looking for similarities rather than differences, and that was dangerous. Since the sixteen points were scrapped no numerical standard has existed. But other experts very seldom challenge the individual decisions of fingerprint officers.

Catherine Tweedy is one of the very few people alive today whose job it is to challenge fingerprint officers. On first impressions, she seems like the kind of teacher kids love because she brings out the best in them – interested, encouraging, knowledgeable. But five minutes in her company reveals something else: a steely intelligence devoted to ferociously logical argument and a passion for getting things right. She has completed a range of fingerprint courses in the UK and abroad, including 'Advanced Latent Fingerprinting' with the Miami Police in Florida. She currently works as a fingerprint expert for a forensic consultancy based in Durham, mostly working as a defence expert, where it is her job to double-check a proportion of the fingerprint identifications made in the UK – albeit a smaller proportion than she'd like. 'I've been doing this from the mid-1990s,' she says, 'and I've come to it as a scientist. I tear my hair out at the assumption people have made right the way through that it is an absolute rock solid science. It's not a science at all. It's a comparison.' The rhetoric used to support forensic fingerprinting has always been scientific in tone. But Catherine Tweedy has spent twenty years trying to remind people that moving forward along a road to certainty is not the same thing as arriving at it; and that going in reverse is also possible.

In 2006, the year of McKie's settlement, Scotland followed England and Wales in scrapping the 16-point system. The US was already using a non-numerical system at that point, although as with all else about American law, much differs from jurisdiction to jurisdiction. In 2011 the results of a public enquiry into the McKie–Asbury fiasco were published. It put the misidentifications down to 'human error' and not to misconduct on the part of the Strathclyde Police. It recommended that fingerprint evidence should from now on be regarded as 'opinion evidence' not fact, and thus should be treated by courts 'on its merits'.

But this message has not trickled down to all fingerprint officers, says Catherine Tweedy: 'They are not being trained to think that an opinion is an opinion. Once you are trained to see things as facts it is extremely difficult to be pulled back to understand that there are shades of grey. You can't be 100 per cent certain in a lot of cases because you only get a fraction of a fingerprint.'

Even when the fingerprint is correctly matched with an individual, mistakes can sometimes be made when the investigator tries to work out what this means. In one of her first cases, Catherine dealt with Jamie, a 14-year-old boy charged with burgling a house in Northern Ireland. His handprint had been recovered on a windowsill in the bathroom. When she met him, he said he'd never entered the house in his life. Catherine went inside and could see why that might be true. The house was such a stinking mess that it was tough to carry out any kind of exhaustive examination. When she looked at the handprint, she saw that it was a clear match for Jamie's. But if someone had climbed in or out of the bathroom window, they would have left footprints in the bath or sink as well as disturbing all the junk underneath the windowsill. And there was no evidence of that.

The officer on scene hadn't been into the other rooms or looked at the two external doors. Catherine carried out her own examination and she couldn't find any other evidence to link him with the inside of the house.

Fired up by Catherine's work, Jamie's defence team found out that the owners of the burgled house had heartlessly kicked their daughter out of the house and on to the streets on her sixteenth birthday. She had gone to stay with friends for a couple of weeks. Then, when she knew her parents were out shopping, she had returned home with her key, walked through the front door and collected her ghetto blaster, moneybox, some of her clothes and a few of her videos.

When her parents returned home they spotted the missing items and called the police to report a burglary. The investigation had begun and ended with the palm print in the bathroom. No further questions had been asked. When Catherine Tweedy had Jamie's friends questioned they told her that they used to play a game called Pirates round the back of the house. Pirates is a variation on the game of tag where players avoid being tagged 'it' by getting both feet off the ground. Jamie, it transpired, was a good climber. His best trick was to shimmy up the drainpipe and hang one-handed off the bathroom windowsill. Without Catherine's tenacity, his agility might have landed him in jail.

Some fingerprints are lifted in far more horrific contexts. On 11 March 2004, during the peak of the rush hour, ten bombs exploded simultaneously on four commuter trains in Madrid, killing 191 people and wounding 1800. The FBI suspected Al-Qaeda involvement.

The Spanish police discovered an abandoned set of detonator caps inside a plastic bag that bore a single, incomplete finger

Spanish forensic experts search for clues in a blasted train carriage outside Atocha train station. The 2004 terrorist attacks were among the worst in Spain's history

mark. This was run through the FBI database, which showed up twenty possible matches.

One of those possibles was Brandon Mayfield, an American-born lawyer living and practising in Oregon. He was on the FBI fingerprint database because he had served in the US army. But, more significantly in counter-terrorist terms, he had married an Egyptian and converted to Islam. He had defended one of the Portland Seven, a group of men who had tried to go to Afghanistan to fight for the Taliban, albeit in a child custody case. But he also worshipped at the same mosque as them.

The FBI decided that Brandon Mayfield was implicated in the bombing, even though his fingerprints were not an exact match, his passport had expired and they could find no evidence that he had travelled abroad for years. They began watching him and his family.

Despite the Spanish police insisting that the fingerprint evidence should be rejected, the FBI agents tapped Mayfield's phone, broke into his house and office, went though his desk and financial records, examined his computers and tailed him. When Mayfield realised he was under surveillance he panicked, so the FBI detained him to prevent him making a run for it. Two agonisingly long weeks passed before the Spanish police matched the fingerprints to the real culprit, an Algerian man named Ouhane Daoud.

Mayfield sued the US government for wrongful detention, and in 2006 received a formal apology and a $2 million settlement.

The FBI later acknowledged one of the problems in the handling of the Mayfield case was that their fingerprint experts had failed to separate the analysis and comparison stages of their examination. First of all the expert should analyse the mark in detail, describing as many minutiae as she can. Only afterwards should she examine possible matches and carry out a comparison. When analysis and comparison happen simultaneously, experts run the risk of finding matching minutiae because they are looking for them. In the view of Itiel Dror, a cognitive psychologist at University College London, 'The vast majority of fingerprints are not a problem, but even if only 1 per cent are, that's thousands of potential errors each year.'

An American experiment in 2006 showed that even experienced fingerprint experts can be swayed by contextual information. Six experts were shown marks that each one had analysed before. But this time they were given certain details about the case – that the suspect was in police custody at the time the crime was committed, for example, or that the suspect had confessed to the crime. In 17 per cent of these secondary examinations, the experts changed their decision in the direction suggested by the information. In other words, they couldn't divorce themselves from the context and judge objectively. This kind of bias is less likely to occur in the UK, where forensic divisions are separated from other divisions in most police forces.

In spite of the question marks raised by experts such as Catherine Tweedy, courts around the world continue to treat fingerprint as infallible and solitary finger marks still send people to jail. In her popular book *The Forensic Casebook* (2004), N. E. Genge states that 'Examiners don't think in any percentages except 100 and 0.' But Christophe Champod, a Swiss expert in forensic identification, calls for fingerprint evidence to be treated in terms of probabilities – bringing it in line with other forensic disciplines – and that examiners should be free to talk about probable or possible matches. He has also called for the overall importance of fingerprinting to be downgraded: 'Fingerprint evidence should be expressed by fingerprint examiners only as corroborative evidence.'

If forensic science were a family, fingerprinting would be the greedy grandfather, hanging on to the best armchair, trying to exercise the sole right to pass judgement, unaware that the times they are a-changin'. Only when the rest of the family understands that he sometimes gets people and places and anecdotes mixed up can his wisdom be treated with appropriate circumspection; then his contribution to the family can be regarded as a healthy and balanced one.

SEVEN

BLOOD SPATTER AND DNA

'Will all great Neptune's ocean wash this blood
Clean from my hand? No, this my hand will rather
The multitudinous seas incarnadine,
Making the green one red.'

Macbeth, II, ii

Blood. It's the key to life. Without it, we die. It's the thread that runs through history, transferring property and power from one generation to the next. From the earliest times, man has understood blood both as a tribal marker and as an individual blazon. In some societies, inheritance flowed not from father to son, but from father to sister's son, because you could be sure that your sister's son was of the same blood as you. You knew for a fact his grandmother was your mother; you couldn't be certain your own sons shared your blood.

It's also been at the beating heart of crime fiction from the beginning. When Doctor Watson first lays eyes on Sherlock Holmes, he is bent over a table perfecting a test for haemoglobin. Watson's slowness in grasping the test's brilliance makes the consulting detective fume. 'Why, man, it is the most practical medico-legal discovery for years. Don't you see that it gives us an infallible test for bloodstains. Come over here now!' Then he stabs a needle into his own finger and uses the resulting drop of blood to show the test in action.

'Criminal cases are continually hinging upon this one point,' he declares. 'A man is suspected of a crime months perhaps after it has been committed. His linen or clothes are examined, and brownish stains discovered upon them. Are they blood stains, or mud stains, or rust stains, or fruit stains? That is a question which has puzzled many an expert, and why? Because there was no reliable test. Now we have the Sherlock Holmes test, and there will no longer be any difficulty.'

The very title of Arthur Conan Doyle's first novel, *A Study in Scarlet,* comes from Holmes's lecture to Watson on the meaning of detective work. 'There's the scarlet thread of murder running through the colourless skein of life, and our duty is to unravel it, and isolate it, and expose every inch of it.' When, shortly afterwards, the pair discover a 'scarlet thread' beginning in a lonely house off the Brixton Road, Watson is nearly sick at the scene, which seems frankly improbable given he's a medical practitioner who has served in the Afghan Wars. But then, I am a writer whose work features blood and gore and yet I am squeamish about blood.

But back to the book. A man has been stabbed in the side as he lay in bed and the blade has pierced his heart. 'From under the door there curled a little red ribbon of blood, which had meandered across the passage and formed a little pool along the skirting at the other side.' This time, there is no need for his new test; instead Holmes assimilates all the physical evidence in the house, and listens to a policeman's take on the anonymous assassin. 'He must have stayed in the room some little time after the murder, for we found bloodstained water in the basin, where he had washed his hands, and marks on the sheets where he had deliberately wiped his knife.'

Reconstructing past events from blood found at a crime scene is known as Bloodstain Pattern Analysis. Conan Doyle's imagination barely touched the edges of what spilled blood can tell modern experts. Two years before *A Study in Scarlet* was published,

Eduard Piotrowski, an assistant at the Institute for Forensic Medicine in Poland, took the first steps in the discipline when he wrote a paper about interpreting bloodstains to explain the course of violent action itself, 'Concerning the Origin, Shape, Direction and Distribution of the Bloodstains Following Head Wounds Caused by Blows' (1895).

Piotrowski put a live rabbit in front of a paper wall, smashed it over the head with a hammer, and got an artist to paint the gory result. The colour illustrations in the paper are as accurate as they are grisly. He battered other rabbits to death using rocks and hatchets, varying his position and angle of attack to see how it affected the shape and position of the bloodstains. We can't know how he felt during the experiments, but in his paper he expressed a nobility of purpose: 'It is of the highest importance to the field of forensic evidence to give the fullest attention to bloodstains found at the scene of a crime because they can throw light on a murder and provide an explanation for the essential moments of the incident.'

Nevertheless, Piotrowski's pioneering work gained little attention until the mid-twentieth century. In a key case in 1955 a handsome doctor named Samuel Sheppard was convicted of bludgeoning his pregnant wife to death in the bedroom of their home on the shore of Lake Erie in Ohio. He maintained that a 'bushy-haired intruder' had attacked his wife (and the back of his own neck, which was injured in a way that would have been extremely difficult to self-inflict).

At his trial, and again at his retrial in 1966, forensic scientist Paul Kirk of the University of California, Berkeley, testified for the defence: 'When a weapon hits a bloody head, the blood flies out like the spokes of a wheel, radially, in all directions.' Kirk showed photographs to the court of a blank area on the wall to the side of the bed where the killer had stood and battered Mrs Sheppard. 'It is entirely certain,' he said, 'that the murderer received blood on his person, and no portion of his clothing that was exposed could have been exempt from bloodstaining.' When police had first arrived at the house Sheppard was shirtless and in a state of

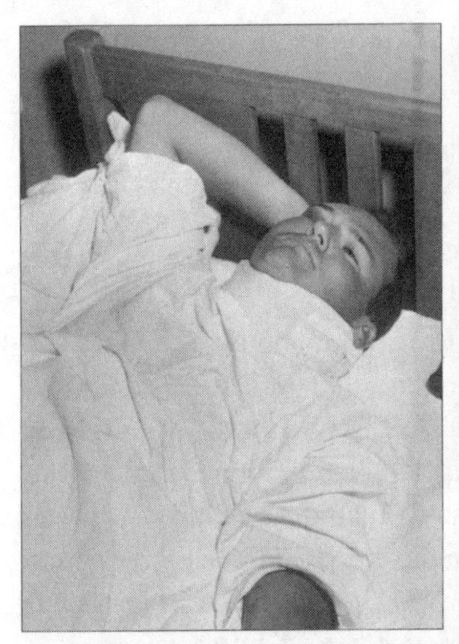

From top left: Samuel Sheppard after the alleged attack, his wife Marilyn Reese Sheppard and Sheppard testifying at his trial in a neck brace. He served ten years of a sentence for second-degree murder; he was found not guilty at a second trial in 1966

shock. The only bloodstain they found on him was on the knee of his trousers. He couldn't remember how he had come to be shirtless: 'Maybe the man I saw needed one. I don't know.' A torn t-shirt of Sheppard's size was later found near the house, with no blood on it. Kirk's convincing testimony at the retrial helped overturn Sheppard's conviction. He walked free after eleven years in jail.

Five years later the US government published the first modern handbook on bloodstain analysis, *Flight Characteristics and Stain Patterns of Human Blood* (1971). The handbook, and its sixty colour photographs, showed CSIs that bloodstaining could reveal how and where a fatal blow was delivered, the kind of weapon used, the likely bloodstaining on the murderer, whether the murderer bled, too, whether they moved the victim post-mortem, or whether the victim themselves moved before dying.

The police still use blood spatter analysis every day: to date, it has helped solve thousands of crimes. But the seismic change in the significance of bloodstains came in the 1980s with the discovery of genetic fingerprinting. The question of 'who' could now be added to the list of 'what', 'where' and 'how'. Since the early twentieth century, scientists had been able to identify the blood type of a suspect from a sample of blood or semen. Though this was useful in narrowing down the pool of potential suspects, the frequency with which some blood types occur in the general population meant it could usually only be used as circumstantial evidence. Blood typing was a far cry from the forensic possibilities offered by DNA.

For thirty-two years Val Tomlinson has been investigating bloodstains at murder scenes, and analysing DNA in laboratories, first with the British Forensic Science Service (FSS) from 1982 until it closed in 2011, and since then with LGC Forensics. She is a mild-mannered, genial woman whose appearance belies her intimate relationship with blood – the way it moves, its inner chemical

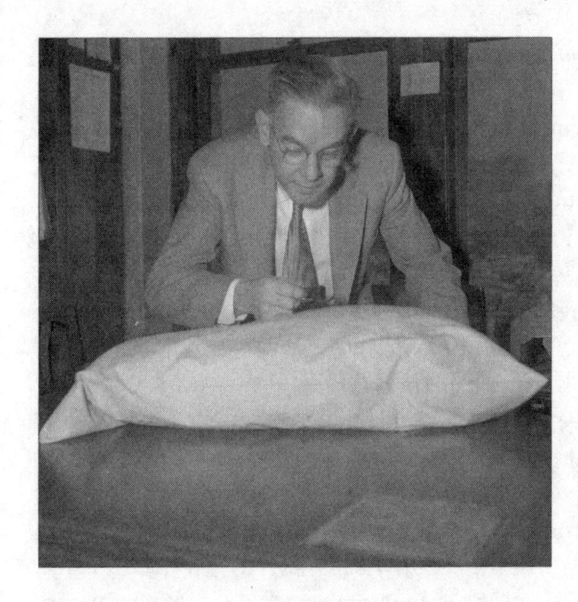

Dr Paul Kirk examines blood spatter on Marilyn Sheppard's pillow

structure, the messages it carries with it – and her profound understanding of the genetic codes that underpin every human life. 'There is a logic to DNA. Scene work is more an art than a science, bizarrely.'

By the time Val arrives at a murder scene with a pad of blank paper under her arm, CSIs have usually photographed and videoed every square inch. 'Many a time I've had a debate with the bobby on the door who goes, "Why are you drawing it, Ms Tomlinson? There's no need."' But – like an artist painting a landscape – Val wants to completely immerse herself in the scene. 'I can take 200 photographs on holiday and when I come home they're just snapshots. But if I stand and draw the scene I'm drawn to specific aspects of it. Very slowly I build up a picture and eliminate irrelevant things. All of the items might be completely irrelevant except one and I can highlight that. A photograph just shows everything that's on the table. There is no emphasis on, say, an item that's turned over, a bloodstained item, a coffee mug.'

Once Val has been in a scene 'for five or six hours' she has ordered it, made it logical. Thus the act of drawing is more important than the drawing itself. 'Even if I haven't got all the answers

I can at least give a discourse about what I've seen and the possible sequence of events.' She imparts this narrative to the Senior Investigating Officer (SIO, the UK's equivalent of a lead detective) and then later to the court, where she uses her scene drawings 'probably just as much as photographs, because the jury can get hold of them, be taken away from all the things that might distract them in the room, and brought into what matters.'

More than anything else at the crime scene, blood matters to Val. Like any other liquid, its dynamics are subject to the laws of physics. If it strikes a floor at right angles it produces a circular stain, often because it has dripped slowly from a person or object. If it travels at an angle it will produce an elliptical stain, usually caused by a punch or a blunt weapon. The longer and thinner the ellipse the more acute the angle of impact. If a group of bloodstains on a surface radiate 'like the spokes of a wheel' they probably came from a blow (or blows) inflicted in one place. A blood spatter expert like Val can calculate the angle of impact of the stains, then attach a piece of string to each one and spool it back at the appropriate angle. The strings will converge at the point where the blow was made. So if, for example, the point of convergence is located close to the floor, the victim could not have been standing when they were struck. Photographs of this 'stringing model' can then be used in court. And, increasingly, angles of spatter impact can also be put into a computer program, such as 'No More Strings', to make a 3D model of blows given at a crime scene.

Cause of death isn't always a mystery: at a scene of battery or stabbing it can be pretty obvious, in which case the SIO might find Val's analysis sheds more light on the incident than the pathologist's post mortem. Is the blood spatter confined to one area, showing that the victim dropped immediately to the floor? Did he stay on his feet and put up a fight, in which case blood might have dripped down his clothes? Did the murderer drag the body for some reason, causing the hair to spread out backwards or the clothing to ruck up, perhaps spreading a trail of blood across the floor? Are the ankles crossed over, indicating that the body has been turned over? The answers to these questions can give the SIO useful information about the actions of the suspect and events surrounding the death of the victim.

Detectives want Val to tell them as quickly as possible how bloodstained the suspect is likely to be. 'The last one I went to, there was an awful lot of bloodstaining at an old Victorian house with lots of rooms. You could see the way the assailant had gone out because every doorway had smears of blood down where the clothing had touched it. Ultimately it turned out they'd burnt the clothing but that was recovered and it was still bloodstained.'

The police are racing against the clock to find the suspect before they dispose of vital evidence. But bloodstaining – like much physical evidence – can be surprisingly hard to get rid of. Val is sometimes called away from the scene of crime to the home of a suspect to examine doors and clothes. 'Often they've had a clean up, so we look at the contents of the washing machine.' Forensic scientists don't give up on evidence easily, something John Gardiner found out to his cost when he tried to dispose of vital evidence in the manslaughter of his wife in 2004 (see p. 174).

But blood analysts can't always report so usefully, especially when they aren't given five or six hours to form an artistic relationship with the crime scene. 'I've heard horror stories of scientists going to scenes and being told, "I want you to look at the blood pattern over there and that's it,"' Val admits. 'To me, that's a disaster waiting to happen. We need to be part of the whole picture.' In some cases analysts testify in court without having visited the scene at all, as happened in a tragic and complicated case that began on 15 February 1997, in the coastal town of Hastings, East Sussex.

In the late afternoon 13-year-old Billie-Jo was putting a lick of paint on the patio doors of her foster parents' home. Siôn Jenkins, her foster father and deputy head of a nearby school, returned from a trip to a local DIY store with two of his own daughters. One girl walked round to the patio to talk to Billie-Jo, and let out a scream. Billie-Jo was lying on her front with her head caved in. Siôn pushed her shoulder up to get a better look at her face, and saw a bloody bubble appear at her nostril, which then popped. He called 999, and paramedics pronounced Billie-Jo dead at the scene.

CSIs found a bloodied metal tent peg near the patio, measuring 46 cm by 1.5 cm. The autopsy showed that the attacker had inflicted at least ten ferocious blows to Billie-Jo's skull. The following day a bloodstain analyst came to examine the scene and found radiating spatter on the wall next to the patio, the inner surface of the patio doors and the dining room floor.

When a child dies in suspicious circumstances, the police often begin to look very carefully at those closest to them. Siôn Jenkins' clothes and the tent peg were sent to the FSS for analysis. On 22 February scientists discovered 158 tiny blood spatters on his trousers, jacket and shoes – too small for the naked eye to see. Were the spatters there because Jenkins had battered his daughter? Or did Billie-Jo breathe a mist of blood on to him with her dying breath?

A number of days after the murder the bloodstain analyst concluded that the blood on Jenkins' clothes was consistent with him being the attacker, but couldn't be certain there wasn't another explanation.

The police arrested Jenkins on 24 February and his trial began on 3 June. A scientist instructed by the prosecution had made bubbles with a blood-filled pipette and burst them next to a white surface. The 'pop' produced a fine spatter that travelled downwards and sideways up to 50 cm – but no spatter rose upwards. Next he filled a pig's head with blood and beat it with the same type of tent peg found near Billie-Jo. This left a fine spatter on his overalls.

A scientist instructed by the defence had done some of his own experiments. He put some of his own blood in his nose and exhaled over a white piece of paper an arm's length away. He also found a fine spatter.

The prosecution argued that Billie-Jo was already dead when Jenkins pushed her shoulder to see her, and so couldn't have made a breath. Paediatrician David Southall testified, 'Anybody approaching a child with an injury who is gasping would be in no doubt whatsoever that the child was breathing and still alive and would report that because it would be so obvious to an observer.' However, neuroscientists had not reached a consensus on exactly when a brain was too injured to cause the respiratory system to produce one more breath. Pathologists for the defence thought that Billie-Jo could have survived long enough to exhale on to her foster father. Under cross-examination the two bloodstain analysts testifying as part of the defence case agreed that the spatter on Jenkins' clothes may have come from the impact of the tent peg.

Siôn Jenkins continued to protest his innocence, but was convicted of murder on 2 July 1998 and was sentenced to life imprisonment. Some rejoiced at the verdict. Others were shocked at how little evidence it was based on, believing that the police had relied too heavily on the assumption that the murderer was likely to come from within the family. In the previous two years there had been eighty-five reports of prowlers and suspicious characters near the Jenkins home in Hastings. The *New Statesman* railed at the conviction, claiming that 'The police had a redhot suspect: someone with a psychiatric history and a known record of violence towards children, whom a number of people saw loitering nearby on the afternoon of the murder. When the police went to interview him he seemed strangely to have disposed of most of his clothing. Whoever the true murderer is, he now has the opportunity, as a result of the vagaries of British justice, to kill someone else's daughter.'

When Siôn Jenkins appealed against his conviction in 2004, the pathologist instructed by the defence presented new evidence about the state of Billie-Jo's lungs. The original autopsy had found them to be hyperinflated, which meant that something (probably blood) was preventing some air escaping. The pathologist suggested that if the blockage was in the upper airways it could suddenly have been released and caused the spatter on his clothes *whether Billie-Jo was dead or alive*. Two retrials followed, both of which ended in the juries failing to reach a verdict, and, in 2006, Jenkins was acquitted. In July 2011 he got a PhD in Criminology from the University of Portsmouth. Now he works with lobbying groups trying, among other things, to ensure that experts who appear in court are properly experienced and impartial. The real murderer of Billie-Jo Jenkins has never been found.

In 1984, Alec Jeffreys was in his lab at the University of Leicester when he experienced a 'eureka moment'. He had been checking X-rays of a DNA experiment comparing members of his technician's family: looking at the results, it was immediately obvious that he had stumbled upon a technique which could reveal the unique variations in the DNA of any individual. Since this chance discovery, DNA profiling (or genetic fingerprinting, as it is sometimes called) has become the 'gold standard' of forensic science. When Sherlock Holmes dreamt up his test for haemoglobin he could proudly state, 'It appears to act as well whether the blood is old or new. Had this test been invented, there are hundreds of men now walking the earth who would long ago have paid the penalty of their crimes.' Within one hundred years of those words being published, real detectives would be able to know *whose* blood they had found at a crime scene. Such knowledge might indicate guilt or, just as importantly, make a compelling case for innocence. For example, if blood found at a rape scene does not belong to the victim or the suspect then, at the very least, you are looking for another person, someone who might have vital information – or who might be the real culprit. In the USA alone 314 people who were languishing in jail, some on Death Row, have been exonerated because of new DNA evidence.

Genetic fingerprinting astonishes people even more than physical fingerprinting did at the turn of the nineteenth century. In the public imagination it stands triumphantly astride other physical evidence. Forensic scientist Angus Marshall remembers 'a legendary case in the States where a jury came back to the judge and said, "We're not going to accept blood splatter evidence, we want to see DNA." They were practically dealing with a confession but they still didn't believe it. It was ludicrous.'

As this suggests, DNA profiling has not always been seen as a purely positive development. But when Alec Jeffreys was asked on the twenty-fifth anniversary of his discovery whether genetic fingerprinting was now being used in a way he was no longer proud of, he replied, 'Catching large numbers of criminals, exonerating the innocent – some of whom have spent more than thirty years in jail – immigrant families reunited ... I would argue the good heavily outweighs the bad.'

To understand the pros and cons of genetic fingerprinting we need to revisit the first crime it helped to solve, in the tranquil and ancient village of Narborough in Leicestershire. On 22 November 1983, the body of 15-year-old Lynda Mann was found strangled and raped near a footpath. She was naked from the waist down and her face was bloodied. Biologists established that a semen sample taken from her body belonged to someone with Type A blood, and a particular type of enzyme secretion, a combination shared by only 10 per cent of men. But with little else to go on, the case went cold.

Three years later, on 31 July 1986, Dawn Ashworth, also aged fifteen, went missing. Her body was found near to Lynda's, just off Ten Pound Lane. Again, she had been strangled, raped and left naked from the waist down.

The main suspect was Richard Buckland, a 17-year-old hospital porter with learning disabilities. Buckland had a troubled past and had been spotted near the scene of the crime. When interviewed, he revealed details about Dawn's murder, and about her body, that weren't publicly available. Before long he confessed to her murder. But he vehemently denied killing Lynda three years earlier.

Convinced that the same man had murdered both girls, police approached Alec Jeffreys at the University of Leicester, five miles from Narborough, who had recently appeared in a local news story about 'genetic fingerprints'. His analysis of the semen samples revealed that the police were right: the same man *had* committed both murders, but he was not Richard Buckland. Despite his confession, Buckland was exonerated – the first person to be proven innocent based on DNA evidence.

The police now had the genetic fingerprint of the killer but they had lost their only suspect. They asked all 5,000 adult men in Narborough and the surrounding villages to volunteer blood or saliva samples. Of the 10 per cent with the particular blood type taken from Lynda and Dawn's bodies, Jeffreys established full DNA profiles. This was a huge and unprecedented undertaking. But, six months and considerable expense later, there was still no match, and again the case went cold.

The following year, a woman sitting in a local pub overheard a local man called Ian Kelly boasting to friends that he had made £200 by posing as his mate, Colin Pitchfork, at the sampling. Pitchfork, a cake decorator – quiet but prone to bouts of temper – had asked Kelly, a colleague at the bakery where they worked, to take the DNA test for him. He said he'd been charged with indecent exposure in the past and wanted to avoid being harassed by the police. The excuse was shaky but £200 was cash enough to stop Kelly asking questions. The woman went to the police, who arrested Pitchfork and took his DNA. It matched. Finally, the detectives had their answer.

In 1988 Pitchfork was sentenced to life imprisonment for both murders. Law enforcement agencies and scientists around the word sat up and took note. Gill Tully was an undergraduate biology student at Cardiff University at the time, and it took her breath away to see such a savage – and seemingly unsolvable – crime exposed by so sophisticated a scientific process. She finished her first degree and then went on to do a PhD at the Forensic Science Service, where she landed a job afterwards. There she was involved in some extraordinary developments at a time of revolutionary change in genetic research. Val Tomlinson had already been at the FSS for six years when Gill arrived, and she recalls the atmosphere of the pre-DNA days:

'It was very hands-on. Personal protective equipment hadn't really been invented. We rarely used gloves. One of the tests for

1 *Crime scene notes taken by John Glaister Junior, the leading forensic investigator in the Buck Ruxton case*

2, 3 *Police officers comb the area where the remains of Isabella Ruxton and her maid, Mary Rogerson, were found. The bodies were recovered in over thirty separate packages, leading many to call the case the 'Jigsaw Murders'*

4 *A maggot's head under a microscope. Note the two prongs, used for scraping decaying flesh into its mouth*

5 *Blowfly (*Sarcophaga nodosa*) on decaying flesh. Blowflies can smell decomposition from over one hundred metres, making them 'the gold standard indicators' of the insect world*

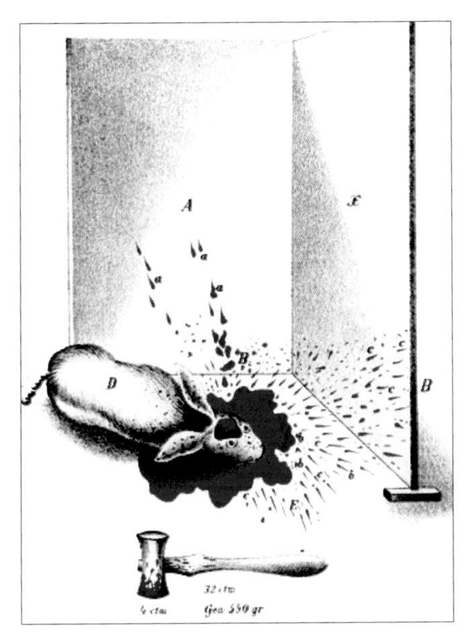

6 An illustration from Eduard Piotrowski's seminal work on bloodstains; as part of his research, he bludgeoned animals with a variety of instruments to observe the effects

7 At the University of Tennessee's 'Body Farm', bodies like this one are left to decompose in a variety of different settings for the purposes of study. This image is part of photographer Sally Mann's series 'What Remains'. Sally Mann, 'Untitled', 2000, gelatin silver print, 30 x 38 inches, edition of three

***8, 9, 10** Graham Coutts, who was convicted of Jane Longhurst's murder, caught on CCTV moving her body from the storage facility where he kept it in the weeks after her death*

11 *Death of a court lady, from a series of eighteenth-century Japanese watercolours depicting the nine stages of a decaying corpse or* kusōzu*: the putrefying body is carrion for scavenging birds and small animals;*

12 *at this stage the flesh has almost all decayed revealing the skeleton. Wistaria blossoms above her body;*

13 *only a few fragments of bone, including the skull and ribs, hand and vertebrae, remain*

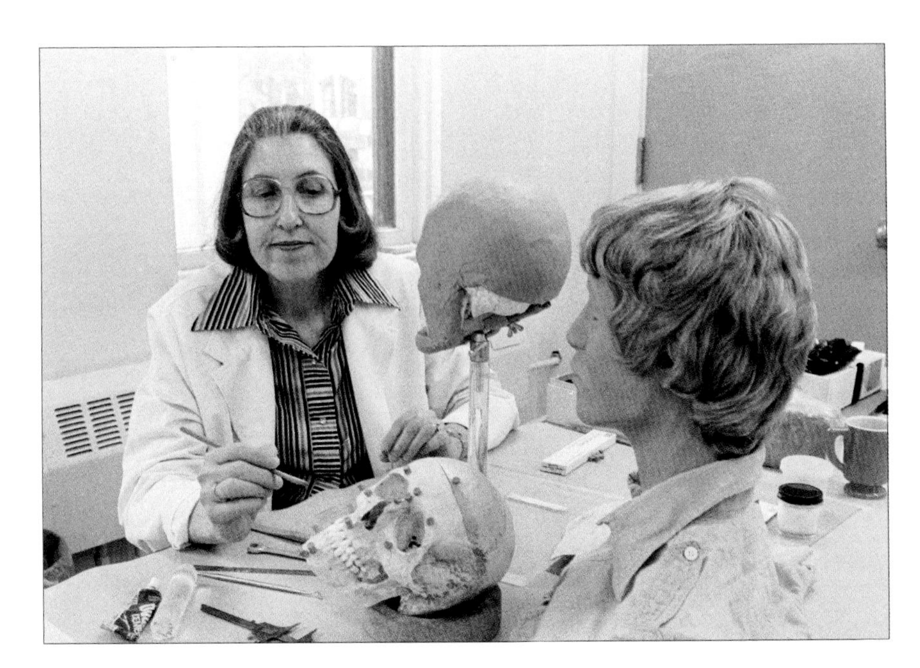

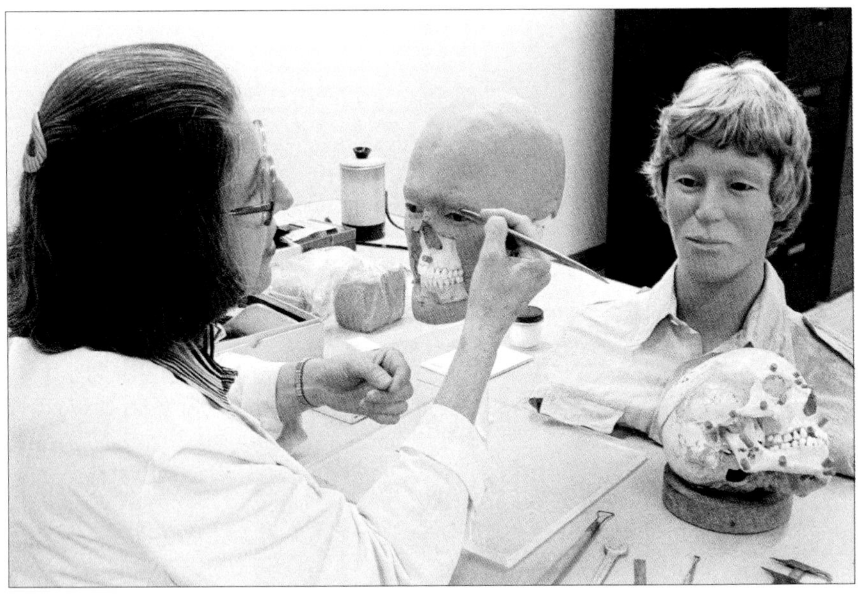

14, 15 *Betty P. Gatliff (see p.198) working on a facial reconstruction of one of serial killer John W. Gacy, Jr.'s nine unidentified victims in July 1980. Photographs of the reconstructed heads were released to the media in an attempt to identify the victims. To her right are a completed reconstruction and a skull with the rubber guides which show the average thickness of tissue on a human face*

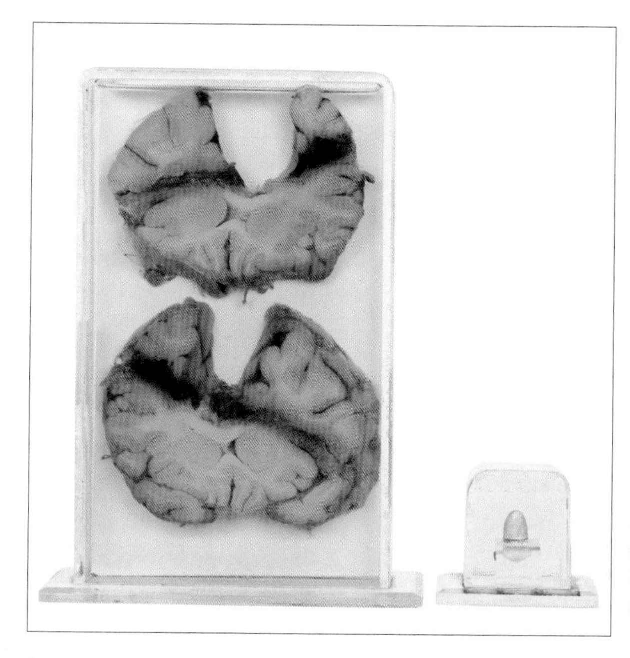

16 *Sections of the brain of a gunshot victim, showing the path of the bullet, and (to right) the bullet itself*

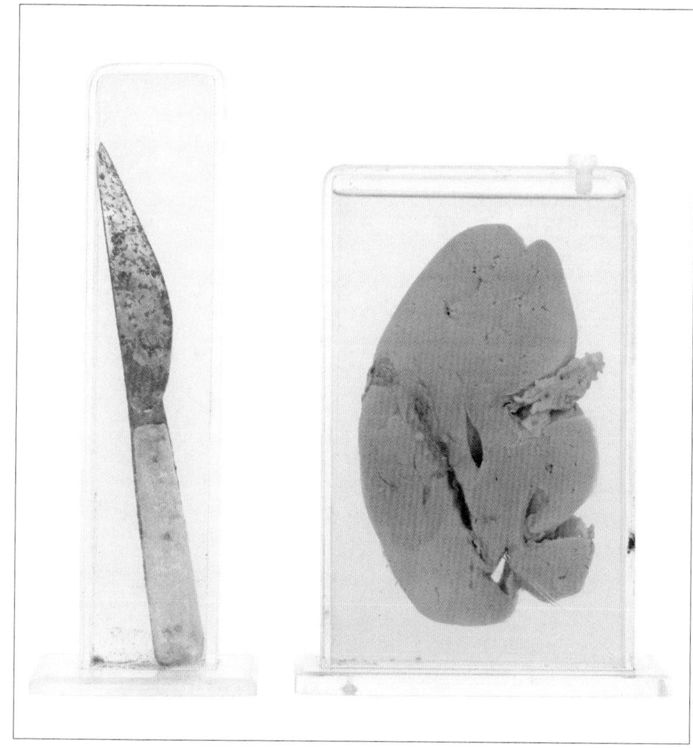

17 *A section from a liver, and (to left) the knife that caused the fatal wound*

18 *One of Frances Glessner Lee's doll's house-sized 'Nutshell Studies of Unexplained Death'. Designed to help train police recruits in detection, the Nutshell Studies depict imaginary crime scenes down to the tiniest details*

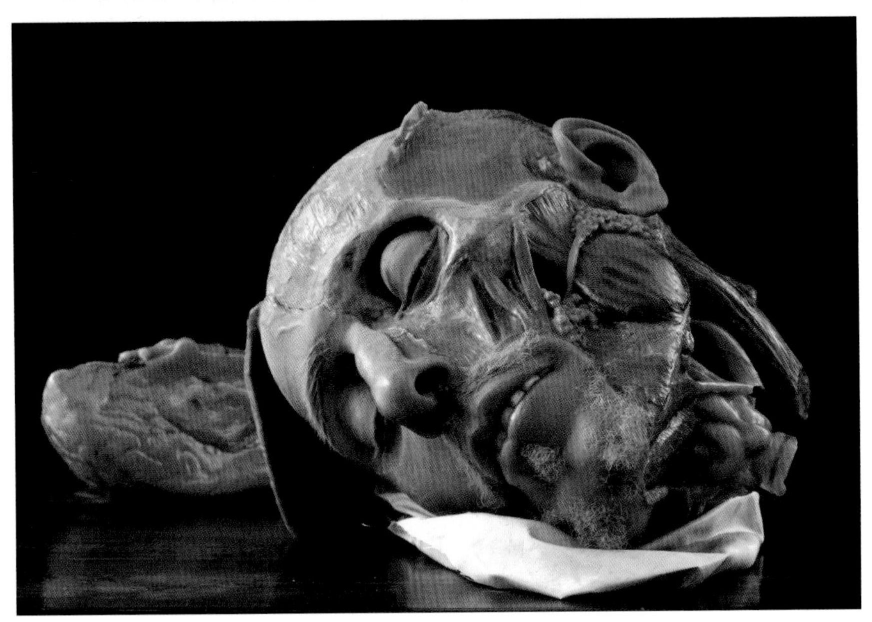

19 *A model of an old man's head in wax, created by the seventeenth-century sculptor Giulio Zumbo. Zumbo created many detailed anatomical models; in this case he built up layers of coloured wax onto a real skull*

Colin Pitchfork, the first person in the UK to be convicted on the basis of DNA evidence

semen staining was whether or not it was stiff to the touch. We didn't have separate offices. Your bench was your office, so you wrote your reports on the same bench where you examined all your dirty knickers and bloodstained items.

'It is quite hilarious to think back about when we started DNA profiling. It was bucket chemistry, literally bucket chemistry, you would make whole vats of salt solutions and you had radioactive substances and you needed a bloodstain the size of a ten pence piece if not bigger to get a DNA profile.

'In the early stages of my career there were no formal training courses other than a very initial one, and then you worked alongside an experienced scientist who took you along with them and you did everything from blood alcohol testing, semen staining, fibres analysis, hairs. I've done pheasant feather cases, I've done salmon poaching cases. I've done leek slashing cases.'

When Gill arrived at the FSS for an internship while she was still an undergraduate at Cardiff, most of the geneticists approached their work with enthusiasm, but without taking much notice of the revolution they were setting in motion. 'At coffee break the main issue was whether there was still a jam donut left,' Gill says now with a rueful laugh. While the Colin Pitchfork case had shown the world how useful DNA could be, she admits, 'We thought it would only be for those occasional really high-profile cases.'

But over the years each innovation has broadened the application of DNA. 'Every time you think, "Oh yes, that would be really good, it's a bit too expensive for routine but that occasional high-profile crime, that could be the thing that makes a difference." And yet a number of those have then become cheap enough and routine enough to use even in burglaries.'

The most significant move away from 'bucket chemistry' was made by Kary Mullis, a Californian surfer and LSD enthusiast who went on to win the Nobel Prize in Chemistry. In 1983 Mullis was driving along Highway 128 when he had a revelation. If he added an enzyme called polymerase to DNA it would, in his words, 'reproduce the hell out of itself'. Using the polymerase chain reaction (PCR), Mullis could take a very small amount of DNA and make it significant enough to interpret. Before long, scientists were using PCR to understand criminal cases that had been cold for up to seventy years, as well as the genealogy of fossilised dinosaurs and buried royalty, and the diagnosis of hereditary diseases.

When Gill Tully started working at the FSS she and her supervisor were the only two people working on refining and using PCR; she regards herself as being 'deeply privileged to have been there from the beginning'. Traditional genetic fingerprinting had relied on bodily fluids and hair, but by 1999 the team that Gill was part of was using PCR to develop a much more sensitive method, known as 'low copy number (LCN) DNA profiling'. To get an LCN profile they needed only a few cells from a potential suspect. Whether it was a speck of dead skin, the sweat from a fingerprint or the dried saliva from a postage stamp, the required amount of bodily substance had spiralled down from the size of a ten pence piece to one millionth of a grain of salt.

LCN profiling has had a seismic impact on the way crimes are investigated. But its road to acceptance has been a long one. Controversial trials involving LCN DNA evidence have provoked reactions from judges and commentators that have forced forensic geneticists to defend and redefine their methods.

One particularly contentious trial that helped to shape the role of LCN DNA in court was triggered by a massive bomb in a small town in Northern Ireland. In 1998, the Good Friday Peace Agreement was reached, supposedly marking the end of hostilities between Unionist and Republican paramilitary organisations. But on 15 August the Real Irish Republican Army, a radical splinter group who parted company with the IRA after the Good Friday peace agreement, detonated a bomb which ripped through a busy street in Omagh, County Tyrone. Responding to a phoned-in warning from the bombers naming the local courthouse as the location, the police had actually moved people into the path of the blast, located in the centre of the town. Twenty-nine people were murdered, including several children and a pair of unborn twins. Over 200 were injured. The Secretary of State for Northern Ireland at the time, Mo Mowlam, described it as 'mass murder'.

Three years later, Colm Murphy, a building contractor, was convicted of causing the explosion and sentenced to fourteen years in prison. It was to be the start of a long and painfully unresolved judicial process. In 2005 his conviction was overturned when it emerged that police had forged notes of interviews they had conducted with him. The following year, the police arrested Colm Murphy's nephew, an electrician called Sean Hoey. At his trial the prosecution's case rested on LCN DNA found on the bomb timers used in the attacks, which a forensic geneticist said was one billion times more likely to belong to Sean Hoey than to an unknown individual. But in the absence of eyewitness testimony, or any other compelling evidence, the case fell apart.

When Mr Justice Weir delivered his judgment on 20 December 2007, he criticised the way the prosecution had made LCN DNA the crux of their case, rather than using it as a guide towards finding other substantial evidence. He complained about the 'slapdash approach' of the police and some forensic experts. He even alleged that the evidence had been 'beefed up' by police, who were guilty of 'deliberate and calculated deception' in their efforts to secure a conviction. He pointed out that the only published papers validating LCN DNA profiling were written by its inventors at the FSS. Ultimately Weir found the method too novel and recommended an urgent review of its use – a bad end to an investigation that cost the state £16 million.

The day after Weir's verdict the Crown Prosecution Service (CPS) suspended LCN DNA profiling, and commissioned a review of its fitness for purpose. Since 1999 it had been used in 21,000 serious crime cases in the UK and abroad – particularly in cold cases. The CPS ordered all live cases involving LCN DNA to be re-examined. One of them involved the brothers David and Terry Reed of Teesside in the north-east of England.

On 12 October 2006 a friend of the former boxer and hard man Peter Hoe received a 4-minute-long voicemail message of the New Age music of Mike Oldfield. But when he played it back and listened carefully he could hear the muffled groans of Hoe, who was bleeding to death from five deep stab wounds in the living room of his house in Eston, near Middlesbrough. The police arrested and charged prime suspects David and Terry Reed. David, the older brother, was known to be jealous of Hoe's tough reputation, and in court Hoe's brother claimed that the attack was a retaliation for a pub scuffle some days earlier: 'They went up to my brother's house and murdered him because David couldn't handle the hiding.'

When Val examined Peter Hoe's living room she found nothing to indicate that the perpetrators themselves had bled, but noticed two small pieces of plastic. 'We see it all the time when knives are used in stabbings. The vibrations and forces project down the knife blade and hit the hilt with such force that it breaks.' Back at the laboratory Val looked more closely at the plastic pieces and, based on her experience, decided they had originated from cheap knives. Traces of DNA were found on them. LCN profiling revealed that it matched the Reed brothers'.

At the trial the defence called an eminent professor of plastics, 'a lovely gentleman from Newcastle University', who had been to Argos and bought a cheap knife with a plastic handle. Then he had put the knife into a machine that bent it slowly until the hilt broke. He explained in court that he'd measured the forces, and was satisfied that a human wrist was not capable of producing them. He declared that the pieces of plastic were unlikely to have come from a stabbing. 'I was sitting in court listening to this,' recalls Val, 'and it was just fundamentally wrong. We had another murder in the laboratory at the same time with four knives. Three of the four had broken in exactly the same way at the hilt.'

The plastics expert had investigated a dynamic event of life and death, steel on bone, plastic on flesh, in a controlled but unrealistic laboratory environment. For Val, that is a scenario fraught with problems. 'Murder is not a replicable experiment. Every one is unique.'

All the time maintaining their innocence, the Reed brothers were both sentenced to a minimum of eighteen years' imprisonment. As they were led from the court, they grinned and thanked the judge while Hoe's mother Maureen cried in the public gallery.

Not long after their conviction, Justice Weir acquitted Sean Hoey of the Omagh Bombing and LCN profiling was put under intense scrutiny. Though its use was reapproved by the CPS in January 2008, enough doubt had been sown that, on 20 October 2009, the Reeds appeared at the Court of Appeal. Their lawyer argued that Val Tomlinson had overstepped the mark when she speculated at the original trial on how the Reed brothers' DNA got on to the pieces of plastic recovered from the scene.

At the appeal of the Reed brothers, in October 2009, the court heard from Bruce Budowle, a former FBI forensic scientist. Budowle argued that LCN DNA profiling was inherently flawed and that its results were not always reproducible. 'The confidence in it has not been assessed,' he said. He accepted that the bits of plastic came from the knives of the murderers, but the Reeds' DNA could have been the result of secondary transfer – that is, they could have touched someone who then touched the knives.

As well as knowing the latest research-based theories, forensic scientists like Val must draw on a database of professional experience to understand what they see. Gill Tully says, 'There have been some interesting judgments from the Court of Appeal in recent years which have really pointed forensic scientists to give opinions informed by experience rather than giving statistical evaluations, which is slightly bizarre for the scientist, although you can see where Their Lordships are coming from.' But, as Sherlock Holmes knew way back when, 'There is a strong family resemblance about misdeeds, and if you have all the details of a thousand at your finger ends, it is odd if you can't unravel the thousand and first.' Val's testimony, both about the shattering of the knife handles and the DNA traces they contained, was based on years of experience with evidence. It was data and opinion; art and science. And, ultimately, the court believed her: while the review made some recommendations about external validation, ultimately it found the method robust and reliable. The three judges at the Reed brothers' appeal decided that the circumstantial evidence was powerful enough to make doubt unreasonable, and upheld their convictions. They thought that Val's professional opinion on how the DNA had got on to the plastic was 'not only possible … but essential'.

The case against the Reed brothers had solid corroborating evidence – such as the fact that Peter Hoe had inflamed the pride of David Reed by knocking him to the floor with a light punch in a pub a fortnight before the murder – as opposed to the case against Sean Hoey, which relied nearly solely on LCN DNA. Valuable lessons had been learnt about the place of DNA in criminal investigations as a key component of a case, but nevertheless, only a component. More such lessons were to come.

In 2011, a woman was brutally raped in Plant Hill Park, Manchester. DNA taken from a swab of the victim linked the crime to Adam Scott, a 19-year-old from Plymouth, who was duly arrested. He was incarcerated in a special segregation wing for rapists and paedophiles, and verbally abused by inmates. But he was adamant that he'd been hundreds of miles away in Plymouth on the night of the crime, and had never even set foot in Manchester.

After four and a half months in jail it emerged that Adam Scott had been the unfortunate victim of laboratory cross-contamination. Some months previously he had been involved in a 'spitting incident' in Exeter, after which police took a swab of his saliva. Scientists placed the swab in a tray at the LGC Forensics laboratory, which was re-used for the swab of the Manchester rape victim. Scott's mobile phone records confirmed that his phone had been in Plymouth when the rape took place.

Andrew Rennision from the government's Forensic Science Regulator said, 'The contamination was the result of human error by a technician who failed to follow basic procedures for the disposal of plastic trays used as part of a validated DNA extraction process.' Adam Scott's case echoed the strange case of the 'Phantom of Heilbronn', a seemingly superhuman female serial killer whose DNA was found at the scene of robberies and murders across Austria, France and Germany in the 1990s and 2000s. In 2009, when the DNA appeared on the burned body of a male asylum seeker in Germany, the authorities concluded that the 'phantom' was simply the result of laboratory contamination: the cotton swabs used for DNA collection were not certified for the purpose, and were eventually traced to the same factory, which employed several Eastern European women who fitted the DNA profile of the 'phantom'.

As with a real fingerprint, a genetic fingerprint should not be enough to secure a conviction on its own. According to Gill, 'DNA doesn't lie. It's an exceptionally good lead and exceptionally strong evidence but there is human interaction in the process [of profiling]. So the error rate is exceptionally low but it's not zero … DNA shouldn't be a lazy way to not do an investigation.'

If in some cases DNA has become a crutch for the police to lean on, in many more it has opened up outlets for their energy, giving them the chance to solve cases both new and old. If DNA found at a crime scene doesn't produce a perfect match when it's run through the national database, it's no longer the end of the line. Because blood tells more than just one person's story.

Familial DNA searching was developed at the FSS by Jonathan Whitaker when he re-examined a grim cold case. In 1973 three 16-year-old girls had been raped, strangled and dumped in woods near Port Talbot in South Wales. After an exhaustive investigation into 200 suspects, police had made no arrests. Then, in 2000, Whitaker used the 28-year-old crime scene samples to develop a DNA profile of the suspect. He ran the profile through the national database, and turned up a blank. Then, a year later, he was struck by an interesting idea. Could there be a family member on the database with a similar profile? He sought permission to search and found a profile with a 50 per cent match. The offender was on the database for car theft, but Jonathan Whitaker was convinced his family tree housed a far more heinous offender. Joseph Kappen, the car thief's father, who had died from lung cancer ten years previously, became the prime suspect. An exhumation order was granted and Whitaker was able to analyse DNA from his teeth and femur. It matched and, although the criminal could not be punished, the triple murder was finally solved.

The first live case to be cracked using familial searching came in 2004. Michael Little was driving his truck under a motorway overpass when someone threw a brick from overhead. It crashed through the windscreen and struck Little in the chest. He managed to steer his truck on to the hard shoulder before succumbing to a fatal heart attack. When scientists fed the LCN DNA from the brick into the database, it produced no direct match, but a familial connection led them to Craig Harman, who admitted his crime and was sentenced to six years for manslaughter. For Detective Chief Inspector Graham Hill of Surrey Police, there was only one reason a conviction was secured: 'There is no doubt in my mind that without this groundbreaking technique this crime would have remained undetected.'

In the aftermath of the Harman conviction Alec Jeffreys said that familial DNA searches raised 'potentially rather thorny' civil liberty issues. The response has to be proportionate to the crime, striking the right balance between an individual's civil rights and the need to identify a perpetrator. Familial DNA searching for forensic purposes remains illegal in most countries. In the US it is only allowed in California and Colorado, though a familial search on DNA extracted from a discarded piece of pizza may have helped to find the 'Grim Sleeper', a serial killer and rapist who terrorised Los Angeles from the late 1980s to the early 2000s. In the UK it's only used for investigations into murder and rape. Since the Harman conviction it has led the police to a suspect in fifty-four serious crimes – producing thirty-eight convictions.

Ethical issues persist. Troy Duster, a sociologist at New York University, points out that because incarceration rates in the US are eight times higher for black people than for white people (for socio-political reasons, including alleged racism on the part of the authorities), familial searches are much more likely to help convict black criminals. The profiles of around two in five of the black men in the UK are on the national DNA database, compared to around one in ten white men. In the US, around 40 per cent of the DNA profiles in the federal database are African-Americans, who make up about 12 per cent of the national population. It is predicted that the DNA profiles of Latinos (about 13 per cent of the population) will soon show a similar skew, mainly due to enforcement crackdowns on immigration violations.

One way to gradually level the playing field would be to profile *everybody*. Already the UK national DNA database has over 6 million profiles, a higher proportion (10 per cent) of its citizens than any other country in the world. DNA from everyone arrested (whether convicted of a crime or not) was held on the database indefinitely until a decision by the European Court of Human Rights in 2008 forced a change. In 2012–13 profiles of 1.7 million innocent people were deleted from the database. Alec Jeffreys had called for this in 2009: 'My view is very simple ... innocent people do not belong on the database. Branding them as future criminals is not a proportionate response in the fight against crime.'

Because so many crimes are committed by repeat offenders, the national database is a powerful police tool. In 2013, 61 per cent of DNA profiles found at crime scenes found matches in the database. The Home Office doesn't record how many of these matches led to a conviction, but it's a formidable help to the police forces, some of whom have advocated mandatory profiling. But others believe that would lead to more false assumptions. DNA from several people is often present at one crime scene for perfectly innocent reasons, particularly since scientists can now produce results from such minute quantities.

This nightmare scenario, along with issues of personal privacy and the huge bureaucratic cost of profiling 60 million people, is probably enough to put the matter to rest for now. In addition, some worry that mandatory profiling would make it easier for criminals to frame innocent people. A defence barrister once put the framing idea to Val Tomlinson in court, claiming that his client's LCN DNA had been planted at the scene by an anonymous other. To prove it, he asked Val a hypothetical question:

'If you were going to set somebody up for this how would you go about doing it?'

'I don't think I could,' said Val.

In Val's experience most set-ups fall down on basic points. 'Children go over the top when they are trying to cover up for their mistakes. And you tend to find that people who frame others distribute too much blood in the wrong way, or a whole bucketful of glass instead of the two little pieces which is what you would expect to persist on a piece of clothing recovered a week after a crime.' Like any powerful tool, DNA can be misused. But, as always, the analysis of evidence is not simply about the collection of data – whose DNA is or isn't there – but also about the interpretative skills of the scientists who deal with it. This is what should – and mostly does – protect the innocent.

Of course, not all criminals want to hide their identities: when political fighters or terrorists commit crimes, they want the world to know who did it. In the Madrid Train Bombings (see p. 34), DNA and politics were central to the case from the very beginning. The timing of the attack, three days before the general election, was significant. In the immediate wake of the bombings, the incumbent government claimed that evidence had been found that implicated the Basque separatist group ETA, perhaps hoping to quash speculation that the bombings were a result of Spanish involvement in the Iraq war. But three days later the self-proclaimed 'military spokesman for Al-Qaeda in Europe', Abu Dujana Al-Afghan, declared responsibility. 'This is a response to the crimes that you have caused in the world, and specifically in Iraq and Afghanistan … You love life and we love death.'

A month later, seven suspects on the verge of a police raid detonated bombs in their apartment, wiping out four of them and one police officer. Scientists could not match the LCN DNA found at the scene (including on a toothbrush), and from other locations, with profiles on national databases. A judge ruled that the scientists should use the DNA to determine whether the suspects who were still at large were of North African or European descent. This would help investigators finally settle whether their targets were members of Al-Qaeda or ETA.

But intermarriage between southern Europeans and northern Africans on both sides of the Mediterranean made differentiating between the two all but impossible using the technology current at the time. Forensic geneticist Christopher Phillips developed a new technique, and was able to conclude that one DNA profile, which did not belong to any of the dead or arrested men, 'almost certainly' belonged to a North African. Familial DNA searching later indicated it belonged to Ouhane Daoud, an Algerian whose fingerprint was also found on unused detonator caps in a Renault Kangoo near the site of the bombings.

While conducting his research into ethnicity, Christopher Phillips was also able to deduce 'with around 90% predictability' that DNA from a scarf found in a van used in the bombings belonged to someone with blue eyes. More and more, scientists can discern details about a suspect's physical appearance from their DNA: traces left at a crime scene can describe the people who were there almost as accurately as an eyewitness.

It all started with ginger hair. In the early 2000s scientists at the FSS found that if a gene (the melanocortin 4 receptor) is switched off in both parents then the child will have red hair. Gill Tully is cautious about the ethical implications of DNA profiling in this way, but overall, she says, 'It's about using things in the right way: when we were developing the red hair test, we had some detectives in Scotland phone us and say, "There's been a shooting and we know from the ballistics what window the shot was fired from. Around there we found some cigarette butts and got a DNA profile from them. We also have an eyewitness account saying a red-haired man ran away from the building. So before we start doing a mass DNA screen of individuals to see if we can find the person who smoked the cigarette butts, can you tell us if they were smoked by a red-haired man?" We weren't quite able to do it at that stage but it was quite a nice example of how these things can be used in an ethical and appropriate way to help direct an investigation so that you don't spend lots and lots of money analysing cigarette butts that are entirely irrelevant and were smoked by somebody months ago.'

Genetic fingerprinting is a powerful indicator of guilt or innocence; the single biggest advance in forensic science since William Herschel and Henry Faulds developed fingerprinting a century before. Much of forensic science is based on subjective interpretation: as explored in the Fingerprinting chapter of this book, experts are sometimes good at finding patterns where they wish to see them, like all human beings. That is a useful skill for a forensic investigator, as long as its intuitive nature is recognised and expressed in court.

Although human error can always creep in, in its simplest form, DNA pulls us out of the trap of subjective bias, interpreting empirical data using objective probabilities that have been refined for over thirty years. When Gill has unmixed DNA from a crime scene which she matches to a suspect she can safely tell jurors that the 'probability of observing that profile if it was from someone other than the person of interest would be one in a billion. That's a conservative estimate that an average juror can get their head around. If you start going into trillions it means nothing.' But life – and crime scenes – are rarely simple. Where, as Gill points out, 'you have DNA mixed from two people, which is often the case, then you need to do a more thorough evaluation of strength of evidence and look at the probability of observing that particular set of mixed peaks both if the prosecution hypothesis is true, and if the defence hypothesis is true.'

There remains a lot more for forensic scientists to learn from DNA. At the moment Val and Gill look at much less than 1 per cent of a person's DNA to judge whether it matches a profile on the national database. As it get quicker and cheaper, 'you could theoretically analyse someone's entire genome'. The possibilities are endless, 'but there are very, very significant ethical and practical issues to answer before you would want to do that. You certainly don't want to be using forensic samples to generate information about people's predisposition to commit crime.' That's a profoundly disquieting thought. We already know, for instance, about the existence of a 'warrior gene' – present mainly in men – which is linked with violent and impulsive behaviour under stress. We do not, in the twenty-first century, want to return to Cesare Lombroso's nineteenth-century *uomo delinquente*, or 'criminal man', or the Victorian discipline of phrenology, which diagnosed a predisposition to criminality from lumps on the skull. A nightmare scenario by any measure.

But, if used proportionately, the future of the genetic fingerprint is more exciting than scary. There are now instruments that can analyse DNA in less than an hour and a half, making it possible to run the profile of an arrested suspect through the national database before they are released from custody. If the search produces a match to profiles found at scenes of unsolved crimes, police have halted a serial offender. Gill explains, 'Habitual burglars, if they are caught, sometimes know the DNA is going to get them so when they are out on bail they will commit more crimes to look out for their family while they are away. Then they'll ask all those crimes to be taken into account and they'll serve their sentences concurrently. There are a few really key cases where it could have potentially prevented serious crime, where people had come into police custody then gone out and committed a serious crime. Whereas if the police had had a DNA result quickly they would never have been bailed.'

At the moment analysing the minute quantities of DNA usually found at crime scenes takes quite a bit longer than an hour and a half, but 'the time will definitely come, and it won't be that far away, when you'll be able to identify a suspect, and not only identify them but potentially go around to their address before they have fenced off all the gear that they've just stolen. It could then be returned to people, things of sentimental value and so on. The potential to do this really, really quickly is not that far away. It's not going to be long.' Let the burglar beware.

EIGHT

ANTHROPOLOGY

'I have seen many strange things, but where shall be seen a thing stranger than this? ... Two lusty porters brought to the witness stand sundry big boxes containing the mortal remains of the woman: they were packed in jars, cigar boxes, paper boxes, tin pails; there were fragments of dry bones, fibers steeped in grisly solutions; anomalous dung and granules, pieces of rag and cloth ... but there, all the time, sat grave professors in the witness chair, interpreting and recounting until, as you listened, the dry bones and dust took on form and life; the rags grew into garments, the garments were fitted on the figure.'

Julian Hawthorne on the Leutgert Murder Case of 1857

We're all fascinated by what forensic science can do. It makes for seductive crime fiction and thrilling TV series. But sometimes we get so caught up in the glamour of the storytelling that we lose sight of the enormity of the crimes that confront investigators in the field. No group of scientists faces that stark reality more than forensic anthropologists. Bloody wars and natural disasters are their front line; bringing home the dead is their vocation.

Kosovo, 1997. As the twentieth century came to a close, one of its most vicious conflicts ripped the Balkans apart along ethnic and religious lines. Each side demonised the other, seeing the enemy as subhuman, as vermin that had to be cleansed to make the land pure again. It's a mindset that inevitably leads to atrocities, and there was no shortage of those in that time and place. I've spoken to some of the investigators who arrived in Kosovo after the war was over; the shadow of things they still cannot speak of lurks in their eyes.

Picture this. A tractor and trailer were making their way down from the Kosovan hills. At the wheel, a farmer who had decided the fighting was growing too close for comfort. In the trailer, all eleven members of his family. His eight children, aged from one to fourteen, squeezed in beside their mother, their grandmother and their aunt. The weather was fine and clear and, in spite of the fear that had become a permanent part of their lives, the family were talking quietly among themselves.

But their attempt to flee to safety had put them in harm's way. Somewhere close by, an enemy lay in wait with one of the most lethal weapons people can face on the battlefield – a rocket-propelled grenade launcher. A child can learn to use it in an afternoon; there are YouTube videos that demonstrate that. It's cheap, effective, highly portable and lethal. It's an icon of asymmetric warfare, the mainstay of guerrilla wars since Vietnam. It often obliterates its targets completely.

Out of nowhere, a grenade hurtled towards the family and exploded, destroying the trailer and annihilating all but one of the passengers. The farmer survived, though one of his legs was wounded in the blast. Shocked and desperate, he dragged himself out of the firing line. Later, under cover of darkness, he crawled around the site of the explosion searching for the other eleven members of his family, collecting as many bloody and broken body parts as he could find. A devout Muslim, he was driven by the need to bury his family as soon as he could. Somehow, in spite of his grief and trauma, he managed to dig a shallow grave and put their remains in the ground.

Eighteen months later, forensic anthropologist Sue Black arrived in Kosovo with the British forensic team to collect evidence for the UN International Criminal Tribunal for the Former

Forensic anthropologists excavating a mass grave in Kosovo

Yugoslavia in The Hague, the first international war crimes trial since Nuremberg and Tokyo in 1945–8. So far, 161 people have been indicted by the tribunal. Seventy-four have been sentenced, and twenty are still being tried. The former president of Yugoslavia, Slobodan Milošević, died in 2006, before he could be sentenced for crimes against humanity. The role of the British team in Kosovo was to exhume mass graves and investigate acts of genocide.

When Sue met the farmer she found him 'the quietest, most dignified man I've ever met'. Sue and her colleagues were seeking key evidence of an unprovoked attack on the people in the trailer. But distant courtrooms in the Netherlands meant little to the bereft farmer. He wanted to mourn his family properly. He thanked the team for coming to exhume the remains of his family, explaining his pain that Allah couldn't find the individual members while they were lumped together in a common grave. He asked her to dig up their remains and bring him back eleven body bags, so he could bury each one separately.

He couldn't have known it, but he had at his service one of the world's leading experts on the bones of children. Sue sent everyone away except for the X-ray technician and the photographer and laid twelve sheets by the makeshift grave. 'We needed the twelfth because I knew there would be elements I couldn't identify with certainty. I was also aware that it would have been very tempting to just put a little bit of something in each of the bags and the father would have been appeased. Of course, that would have been completely and utterly morally incorrect. But more importantly it also would have been judicially unacceptable. We are there for forensic purposes, not humanitarian. Our job is to collect evidence, analyse evidence, present evidence and, when we go into court, to be able to justify what we have done.' She imagined a defence expert opening one of the body bags and finding that the material was not what it purported to be. It would completely discredit the prosecution.

So she set to work. After eighteen months, decomposition had done its work and most of the material she had to deal with was bone. The adults were relatively easy to distinguish from one another because they were bigger and there were fewer of them. The eight children were much harder. Sue painstakingly separated the fragments. After several hours she had identified the six youngest children. All that remained were two sets of upper limbs, which had belonged to 14-year-old twin boys. 'There was nothing else of them. Just humeri and clavicles. But one of the sets of upper limbs was attached to a Mickey Mouse vest. I said to a police officer, "Go and ask the dad which of his children liked Mickey Mouse. Don't say which of the twin boys or anything leading like that. If he comes back with the name of one of the twins, then we can separate them."' The officer came back with the father's reply. He'd named one of the twins. '"He adored Mickey Mouse. That's his vest."' An hour later Sue brought the twelve body bags to him. 'That's what he wanted more than anything. Giving him his family back was the absolute and utter least we could do, considering what he'd been through.'

Sue is director of the Centre for Anatomy and Human Identification at the University of Dundee. At the heart of her job in the field is the recovery and identification of skeletal remains. Are they human? What sex, what age, what height, what ethnicity?

When did death occur? Why? If a corpse is intact and not too decomposed, a pathologist may be able to answer these questions. If not, a forensic anthropologist is needed to analyse not just the bones but all the 'human remains' left behind: hair, clothing, jewellery, any of the many items we collect and carry with us every day. As we will see, even the images we leave behind ourselves on camera or video can be analysed for clues that can take years of experience even to notice. Throughout her working life, Sue has traced the secret patterns of the human body, pioneered extraordinary techniques to uncover people's identities, and taught scores of anatomists, anthropologists and medics how the human body is put together.

The material she has taught her undergraduates, the field trips she has taken them on, and her own research have all been profoundly influenced by her 4-year post-war involvement in Kosovo. Sue describes Kosovo as the big turning point of her career in part because, while she was working there, she was able to share knowledge and experience with a number of national forensic teams. These included the famous Argentine Forensic Anthropology Team, who pioneered the application of their expertise to cases of human rights abuse in the 1970s and early 1980s.

Between 1976 and 1983 Argentina was ruled by a military junta which took violent and repressive action against those it considered left wing or subversive, a conflict named by the perpetrators '*Guerra Sucia*' or 'Dirty War'. In Buenos Aires and other cities, civilians were kidnapped in public areas or snatched from their homes and taken to one of 300 secret prisons around the country. Many were brutally tortured – men, women and children alike. Survivors describe being strapped to metal grids and electrocuted. Pregnancy was no deterrent to the captors' cruelty. Others were drugged, blindfolded and dropped out of planes over the River Plate between Argentina and Uruguay, their bodies washing up on both shores. When not deposited in unmarked graves

Clyde Snow testifies at the 1986 trial of nine former Argentinian military junta leaders for murders carried out during Argentina's 'dirty war'. Snow's testimony helped convict six of the defendants

or water, bodies were sent to morgues and marked 'no name'. A worker described 'bodies stored for over 30 days without any sort of refrigeration ... clouds of flies and the floor covered in a layer about 10.5 cm deep in worms and larvae'. As many as 30,000 civilians were victims of the 'Dirty War', and around 10,000 were among the 'disappeared'.

In 1984, after the fall of the junta, local Argentine judges began to demand that bodies be exhumed from unmarked graves and identified, to enable people to establish what had happened to their lost relatives, and to bring their murderers to justice. The local doctors following the judges' orders had little experience of analysing skeletons and desperately needed help. In 1986 Clyde Snow, an experienced forensic anthropologist who had worked on the Kennedy assassination and the victims of serial killer John Wayne Gacy, came from the US to train the founding members of the Argentine Forensic Anthropology Team. 'For the first time in the history of human rights investigations,' explains Snow, 'we began to use a scientific method to investigate violations. We started out small, but it led to a genuine revolution in how human rights violations are investigated. The idea of using science in the human rights area began here, in Argentina, and it is now used throughout the world.'

Snow collected a small but dedicated team of young Argentineans about him, often training them on the job. In the early months, he describes how his students would break down in tears at the site of graves, and he began to drill a 'mantra' into them: 'If you have to cry, you can cry at night.' Once the anthropologists had exhumed and documented a body, investigators tried to match its biological profile with medical and dental records of known missing people. In recent years, anthropologists have extracted DNA from the bones of those still unidentified and linked them to living relatives. By 2000, sixty skeletons had been identified, and a further 300 were still under investigation; a tiny proportion of the whole, but a beginning. One identification was of Liliana Pereyra, who was snatched as she walked home from work on 5 October 1977. She would later be tortured, raped and murdered by her abductors. When Liliana disappeared she was five months pregnant. At the trial of nine military leaders in 1985, Clyde Snow was able to testify to Liliana's identity, telling the court that 'in many ways the skeleton is its own best witness'. Evidence from Liliana Pereyra's bones, along with several other representative skeletons, helped convict six of the defendants.

The Argentine team went on to work in more than thirty countries around the world, exhuming mass graves and training others to conduct their own forensic investigations. They trained the Guatemalan Forensic Anthropology Foundation, which was set up to investigate human rights violations during the thirty-year civil war. They worked with the Truth and Reconciliation Commission in South Africa in the aftermath of apartheid. They also collaborated with a team of Cuban geologists in 1997 to identify the remains of Che Guevara in Bolivia. It was known that he had been shot in the legs, arms and thorax in 1967, and had had his hands cut off by Bolivian soldiers to confirm his identity. Anthropologists

Members of the Argentine Forensic Anthropology Team excavate a common grave in Cordoba province, Argentina, where they found some 100 unidentified bodies thought to be victims of the 'Dirty War'.

searching for his remains found seven bodies in two graves. One of the bodies wore a blue jacket, in the pocket of which the team found a small bag of pipe tobacco, which a Bolivian helicopter pilot had given to Guevara shortly before his death. The identification was confirmed by dental records. Thirty years after his execution, Che Guevara was returned to Cuba to a hero's welcome.

The expertise the Argentinian forensics team were able to share in Kosovo helped others like Sue Black to expand their own knowledge and techniques which have formed the basis of developments in the discipline worldwide. Sue herself has worked in such diverse situations as Sierra Leone, Iraq and Thailand in the aftermath of the 2004 tsunami, as well as running extensive training programmes in the UK.

And still atrocities continue that require her expertise. In January 2014 a Syrian defector, code-named 'Caesar' and said to have been a military police photographer, smuggled out 55,000 photographs showing the bodies of 11,000 men who had allegedly been detained while fighting against the Assad dictatorship. The regime questioned their legitimacy, claiming an opposition group had faked them. Sue was asked to examine the photographs to determine their authenticity. She has described them as 'the worst example of violence I have come across in thirty years of forensic science'. Whilst the Kosovo atrocities mainly involved gunshots, and the tsunami was an act of nature, these photographs revealed systematic torture. The bodies showed signs of starvation, strangulation and electrocution. They had suffered beatings, burnings and eye gougings. Sue was asked whether the evidence of torture was credible and if the deaths warranted further investigation. Her answer to both questions was, 'Emphatically, yes.'

Thankfully, most of the work of the forensic anthropologist doesn't involve investigating torture or genocide. And it's only infrequently that they're called to the scene of what are known as 'mass fatality events' – natural disasters, train crashes, the London Transport bombings of 2005. In fact, most of their cases are on a much smaller scale. But just as significant to those touched by these individual fatalities.

John and Margaret Gardiner lived in Helensburgh on the west coast of Scotland, an hour's drive from Glasgow. John was an ex-merchant seaman whose ability to dream big was matched only by his ability to run up debts. In October 2004 John unveiled his latest get-rich-quick plan: to build luxury kitchens. Margaret was unimpressed by this scheme, and told him so in no uncertain terms.

At her office a few days later, she took a call from a bank clerk telling her that there was a problem with her £50,000 loan application. The news surprised her because she hadn't made a loan application. To the best of her knowledge she'd never asked for a loan in her life. In the course of the conversation, it became clear that John had persuaded another woman to pose as Margaret and fill out the form in her name. This was too much for Margaret. She told her colleagues she was going home to have it out with her husband, and set off to turn him out of the house. That was the last time she was seen alive.

When people asked where his wife was, John reeled off a vague story. But the one thing he couldn't explain away was why she had suddenly stopped phoning her elderly parents, a duty she carried out every single evening. When Margaret's disappearance was reported to the police, they took the matter seriously and sent a forensic team round to the house. In the bathroom the CSIs found some blood on the base of the bath tap. Margaret's blood. They put an endoscope down the U-bend of the bath pipe and discovered a chip of tooth enamel. They checked the washing machine in the kitchen, swabbed around the door and found more of Margaret's blood.

But none of this meant Margaret was dead. She could have stumbled and fallen in the bathroom, chipping her tooth and cutting herself in the process, then put her bloodied clothes in the washing machine. However, the CSIs were determined to do a thorough job. They took the filter out of the washing machine and there they found a tiny cream-coloured fragment, a mere 4 mm wide and 1 cm long. They weren't sure but they thought it might just be bone. They could have ground it into a powder and had it tested for DNA. But fortunately they understood that before you apply a test that destroys the evidence, it's important to apply any available techniques that preserve it.

So they brought the fragment to the Centre for Anatomy and Human Identification, where Sue Black identified it as not just a random piece of bone but the left greater wing of the sphenoid bone. That part of the bone sits in the temple and right underneath it is a crucial branch of a major artery. With that piece of bone missing, Margaret Gardiner would have bled to death. She couldn't possibly be alive.

Such a tiny piece of evidence made a nonsense of John Gardiner's fabrications. Confronted with the incontrovertible fragment, he quickly gave the police a new version of events. Margaret had burst through the front door in a rage, he said. Their slanging match quickly turned physical. Margaret got away from her husband's clutches. He gave chase. She hurried out of the house. Tripped on the top step. Hit her head on the patio. She suffered catastrophic bleeding. John carried his wife into the bathroom, which explained the blood that had been found there. Then he noticed blood on his jumper, so he stuffed it in the washing machine. He washed it on a cold cycle with a non-biological detergent, thus preserving the DNA on the fragment, which must have been trapped in the fibres of his jumper. His story matched the evidence. Later, he told his daughter that he had wrapped Margaret in a sheet and put her in the river. But even though Margaret Gardiner's body has never been found, on the basis of that tiny fragment of bone containing her DNA, her husband was convicted of manslaughter.

Long before the formal science of anthropology was being used in twenty-first century cases like Margaret Gardiner's manslaughter, an interest in bones had played a role in a legal judgment. The case in question concerned a thirteenth-century official whose story was presented in the Chinese coroners' handbook, *The Washing Away of Wrongs* (1247). A man had killed a young boy and seized his possessions. A long time afterwards, the crime was discovered. The criminal confessed, saying he had beaten the boy and thrown him into a lake. The boy's body was found in the lake, but the flesh had rotted away, leaving only the bones. A high-ranking official thought that the bones might belong to someone else. No one dared pass a contradictory judgment, and no inquest could be held.

But some time later, another official reviewed the case records and noticed that a relative had described the boy as 'pigeon-chested'. The official went to look at the skeleton. Sure enough, the victim's ribs met at an acute angle. A new inquest was called for. The murderer's confession was validated and he was finally punished for his crime.

But, despite this early success, it was many centuries before the science of bones was formally introduced to the courtroom. The first recorded instance of an anthropologist featuring in a criminal trial was in 1897 in America. George Dorsey was an ethnographer who specialised in Amerindians, and in 1894 he was the first person to be awarded a PhD in anthropology by Harvard University. He had been taught by Thomas Dwight, the so-called 'father of forensic anthropology', who had led the way in the subject's early development and was able to analyse the variability of human skeletons with unprecedented accuracy. At the time of the trial, Dorsey's passion for collecting artefacts, especially skeletons, took him on expeditions across both North and South America, and he had brought back a large number of Inca mummies from Peru.

In 1897, Dorsey was drawn into a case that occupied the front pages of the newspapers for weeks. Adolph Leutgert had emigrated to Chicago from Germany as a penniless 21-year-old in 1866. Like John Gardiner, he was a man of grand ambitions. Unlike him, he was good with money. For fifteen years he worked odd jobs at tanneries and removal companies until he had saved $4,000 – enough to build a factory and set up the A. L. Leutgert Sausage & Packing Company. Sausages from the factory were soon distributed all over the city and beyond, earning Leutgert the title of 'The Sausage King of Chicago'.

Just before opening the factory the burly sausage entrepreneur had married a petite and attractive woman called Louisa. But the marriage was a long way from the American Dream. Adolph began to sleep with other women. Rumours spread that he beat his wife.

On 1 May 1897 the pair went out for a spring stroll. But only Adolph returned. Unconvinced by Adolph's story that she had run away with another man, Louisa's family went to the police, who carried out an extensive search, eventually turning to the Leutgert sausage factory. An eyewitness told them that he'd seen

Adolph and his wife entering the factory at 10.30 on the night she had disappeared. The night watchman corroborated this. More than that – Mr Leutgert had given him an errand and told him he could take the rest of the night off.

As they walked through the factory, the police noticed a peculiar smell coming from a large vat used for steaming sausages. Peering into the vat, the officers noticed sludge at the bottom, which one described as having a 'very sickening [smell] … something dead around it'. They decided to investigate further.

'A plug on the outside, near the bottom was withdrawn, and some gunny sacks … were spread on the floor at the [plug] hole. As the liquid passed out, a slimy sediment and a number of small pieces of bone were deposited on the sacks. The vat was then further searched and, at the bottom, beside other bone fragments, there were found two plain gold rings, stuck together and covered with a slimy, reddish-gray substance; the smaller was a guard ring, the larger a wedding ring, and on the inner surface of the latter was engraved in script "L.L"' This was later confirmed as Louisa Leutgert's wedding band, a gift from her husband. Inside a furnace, the police also discovered some small pieces of what looked like bone, and a piece of burned corset. In light of this weight of evidence, Leutgert was arrested.

The trial was held at the Cook County courthouse in the summer of the same year, amid an atmosphere of fevered public interest. George Dorsey and some of his colleagues from the Field Museum in Chicago testified for the prosecution. Dorsey said that the bones found in the furnace were human, and that they included bones from the foot, finger, ribcage, toe and skull of a woman. A further witness testified that the slime found in the vat contained hematin, a chemical created by the decomposition of haemoglobin, found in human blood.

Yet another witness said that, before Louisa's disappearance, Adolph had bought several hundred pounds of lye – a caustic compound that can be used for such wildly different purposes as curing meat, cleaning ovens and making methamphetamine – which he had gradually added to the sausage-dipping vat. Adolph testified that the lye was for cleaning the factory. The prosecution countered that lye is strongly alkaline, good at dissolving large objects.

The first trial produced a hung jury; the jurors were so far from agreement that they nearly came to blows in the deliberation room. But Leutgert wasn't off the hook. The following year a retrial was held. George Dorsey testified again. And this time Leutgert was found guilty of murdering his wife.

George Dorsey made a good impression on the witness stand. As the *Chicago Tribune* noted, '[I]t was abundantly evident that his sole concern was to present the exact truth as he knew it, exaggerating naught, and setting naught down in malice ... his knowledge was ... well systematized, so well in hand, so sound, precise and broad.' By contrast, the defence's expert, William H. Allport, humiliated himself on the stand by identifying a bone from a dog as one from a monkey. To muffled laughs from the jury, he prevaricated that 'there is a class of dog monkeys'. But away from the courtroom, Dorsey faced so much harsh criticism from other anatomists about his handling of the case – including the spiteful Allport, who jeered at his 'identifying a woman from four fragments of bone the size of peas' – that he abandoned forensics entirely. But the press coverage had undoubtedly put forensic anthropology on the map for the general public for the first time.

Forensic anthropology in its modern form is a relatively new field. In the early part of the twentieth century the analysis of skeletal remains took slow and gradual steps forward. But move forward it did.

Aleš Hrdlička had been born in Bohemia (now part of the Czech Republic) before emigrating in 1881 at the age of thirteen to the US, where he became obsessively interested in human origins. Like George Dorsey before him, Hrdlička studied the indigenous peoples of America. At thirty years old, he set off on a

5-year expedition across America, studying skeletons as he went. His conclusions led him to an original theory – that East Asian people had travelled across the Bering Strait and colonised America around 12,000 years before. This concept has since become a scientific commonplace, thanks in part to DNA profiling. But, as well as human origins, he was interested in the origins of human evil, and studied the anthropometric characteristics of criminals and 'normal' Americans to discover whether the measurements of wrongdoers were different. By 1939 he was able to announce: 'Crime is not physical, it is mental.'

Hrdlička's expertise had not gone unnoticed. In the 1930s the FBI wondered whether this fledgling science might help them in cold cases, and turned to him for help. Hrdlička was consulted in more than thirty-five FBI cases, determining the identity of skeletal remains, their age and whether there had been foul play. Hrdlička brought a greater organisation and systematic approach to forensic anthropology and, on his death, J. Edgar Hoover, the Director of the FBI, praised his 'outstanding contribution to the science of crime detection'. While he was pursuing these investigations, Hrdlička was also preparing the next generation of forensic anthropologists, teaching students at the Smithsonian Institute.

Just as Sue Black's own professional turning point emerged from her work on the Balkan genocide, some of the biggest breakthroughs for forensic anthropology in the twentieth century came from its most tragic events. One of Hrdlička's most gifted apprentices, T. D. Stewart, worked in a warehouse in the Japanese city of Kokura, identifying the war dead from the Korean War. The task was made particularly difficult because of the effects of modern explosive weapons on the human body. The remains arrived in enormous boxes filled with bones, and the process was an arduous and heartbreaking one. But Stewart seized the opportunity given to him: he had unparalleled access to a huge sample group of human bones. He began a painstaking catalogue of measurements, gradually building up a database that allowed accurate predictions of height, weight and approximate age from skeletal remains.

Another anthropologist who made a vast contribution in this area was Mildred Trotter, who in 1947 had started work at the American Graves Registration Service in Hawaii. Impatient with the data she already had for predicting height and age – which was fifty years old and from France – she began to take her own measurements, using bones from soldiers killed in the Second World War. Today, the US Army Central Identification Laboratory remains the largest human identification laboratory in the world, and Stewart's and Trotter's measurements continue to be used widely.

The lessons learned in Hawaii have spread outwards and informed other forensic anthropologists dedicated to the identification of the dead. Education is at the heart of what Sue Black has pioneered at the Centre for Anatomy and Human Identification (CAHID) in Dundee. In 2008 the centre set up a free 24-hour email service for police, its aim to answer the key question 'Is this bone human or not?' in under ten minutes. The caseload increases in the summer, when people go out into their gardens and dig, or go walking in the countryside.

Answering that vital question can be very difficult. The effects of weather on a landscape and of scavenging animals can scatter and destroy skeletons, sometimes leaving only a single bone behind. The ribs of sheep and deer are very similar to human ribs, and are easily confused. The small bones and teeth of children are also similar to those of animals. And because they are numerous – children have around 800 bones until they fuse into the 209 bones of an adult – they can be easily scattered over a wide area in a rural landscape. (Clyde Snow estimates that a child's skeleton has only forty-six 'findable' bones.)

In 2012 the CAHID service answered a satisfying 365 bone cases. A bone case a day. But how many of those bones turned out to be human? 'Ninety-eight per cent turn out not to be,' explains Sue Black. But even a negative result provides an important service. 'It's about telling police not to start off a murder investigation on Ermintrude the cow, because they won't get very far with it.'

And there will always be the 2 per cent that once belonged to a living, breathing person. And that's where the skills of an anatomist or an anthropologist come to the fore. To identify which human bone they're looking at, they first have to measure its size and thickness, then look at the subtle bumps, grooves and indentations that define the function of each of our bones. Depending on the bone in question, sex can sometimes be determined: men tend to have larger and more robust bones than women. They also have a heart-shaped opening in their pelvis, compared to the circular one through which women give birth. Men's skulls tend to be larger, too, with squarer jaws than women's.

A couple of years ago I was with Sue Black in her office when a uniformed police officer came in carrying a paper bag containing a bone he'd found on the beach near Kirkcaldy, where I grew up. Sue gloved up and took the bone from the bag with a theatrical flourish. We could all see it was a jawbone, a few teeth still doggedly in place. 'It's human, all right,' Sue said solemnly. I was convinced it was a wind-up, staged for my benefit. A human bone from the very beach where I used to play as a child. But no, Sue insisted. She took pity on me and explained, 'There's nothing here to interest the police. This is a very old bone. The person this belonged to has been dead for a very long time. Far too long to have any legal significance. We get that kind of thing all the time.'

An encounter with a stranger's jawbone is something most of us would regard as gruesome. The word isn't in Sue Black's professional vocabulary. Nor are 'disgusting', 'repellent' or 'queasy'. The human body in all its glories and ignominies is her workplace and she brings to it a calm competence that has no place for squeamishness. She maintains that any lingering discomfort she might have had about blood and flesh and bone was dispelled by her first job – working part time in butcher's shop from the age of twelve. She recalls it being so cold there that 'when a lorry with livers in it arrived straight from the abattoir, we used to race each other to the back, so at least we could get our hands warm'. The usual things that push people away from anatomy didn't bother her. But what pulled her to it?

Not, primarily, a desire to deliver justice to criminals. She is a researcher at heart, obsessed with trying to understand the mysteries of the human body. Only later did she see how that understanding might unlock the mysteries of the destruction we humans wreak on each other. As an anatomy undergraduate, and the first member of her family to go to university, Sue Black found dissecting people 'the most enormously humbling experience'. She saw them as people who had offered themselves as corporeal textbooks for scientists to pore over, so that they might make breakthroughs that would benefit others. Sue chose bone identification for her first research project and soon realised how readily it could be applied practically.

Her first case involved identifying a microlight pilot who had crashed off the east coast of Scotland. Sue was apprehensive about how she'd react to seeing the crushed body of the pilot, but, faced with the reality, the necessary clinical detachment slipped into place. She resolved that case and decided that she could make a career in this field.

Sue's work brings her face to face with the sort of questions most of us can relegate to our leisure pursuits. 'We all love a good mystery,' she says. 'We all love a good crime. We all read the books and watch the shows, because we have an innate curiosity about the human body and its anatomy. We can use that curiosity to solve a problem and the problem is "Who is it?" "What is it?" So I get a wonderful combination, where I am doing anatomy which is where I feel most at home; I am applying it to a problem in the world that really needs addressing, and I am satisfying basic human curiosity at the same time.'

At the beginning, Sue Black's forensic work centred around identifying victims. A successful identification often helps to define a crime, and makes a criminal investigation possible. But the investigation of a crime is about much more than who's been on the receiving end. At its heart it's about who perpetrated the act. That's been at the core of crime fiction since the origins of the genre in the nineteenth century. Good scientists, like good detectives, develop new techniques to overcome particular problems. If these techniques are successful they can be applied to other similar cases. For Sue Black the blazing of new trails has always been a driving force. Wherever she can, she is determined to extend the range and scope of forensic anthropology. In recent years she has spent less time working out the identities of victims than she has on nailing the victimisers.

Nick Marsh, Head of Photography for the Metropolitan Police, worked alongside Sue in Kosovo, where they became friends as well as professional confidants. After he returned to the UK, he was confronted with a seemingly hopeless case in his photographic unit. A 14-year-old girl had come to the police alleging that her father was abusing her at night. She had told her mother, who hadn't believed her. The girl knew she needed evidence. Because she was tech-savvy, she knew that a webcam would switch to recording infrared light when it is dark. The girl set up her camera, pointed it towards her bed and clicked 'record'.

She brought the resulting video to the police. The seemingly intractable problem Nick Marsh faced was that he could see there had been abuse. But because the camera had a very narrow view, the face of the perpetrator remained out of shot. Without a face or other obvious identifying marks, the video wouldn't be enough to convict the father.

And so Nick turned to the one person he suspected might be able to help. When she viewed the video, Sue said, 'It was one of the spookiest things that I've ever seen. I felt the hair go up on the back of my neck. At about 4.15 in the morning a pair of legs came into the shot of the camera and stood there. You can see where she is lying on the bed. She is wearing her pyjamas and it is her buttock region that we can see. And he just stands there – and I know it's a 'he' because of the very, very hairy legs – and then very slowly extends his forearm, puts his hand underneath the covers.'

Like Nick, Sue's first thought was that it would be impossible to identify the abuser. But she looked more closely at the footage and noticed that the infrared light had revealed the perpetrator's deoxygenated blood, highlighting the superficial veins on his forearm. She already knew that superficial vein patterns differ widely. The further from the heart, the more clearly differentiated they are, so the veins on the hands and the forearms are the most individual ones that our bodies display. But to identify someone on the basis of these patterns would be a forensic first. At Sue's suggestion, the father's right arm was photographed. The veins matched perfectly with those of the man in the video.

When the case went to court, the defence questioned the admissibility of Sue's evidence. The judge agreed that vein pattern analysis had no track record whatsoever. The jury was cleared out so the defence and prosecution could present their arguments on whether the evidence should be allowed or not. The judge asked Sue what she planned to say. By now she had realised that she should have photographed *both* the father's arms to demonstrate how forearm veins differ, even on the same individual. In a bid to make her point, she asked the judge to turn his own hands over and look at the differences in his own veins. The judge asked her if her evidence proved beyond doubt that the perpetrator was the father. 'No,' she said candidly. 'I haven't done enough research to be sure that the pattern wouldn't match anybody else in the world.' The defence were desperate to get the evidence thrown out. It came down to the judge. Ultimately, the judge deemed the evidence to be admissable based on Sue's anatomical experience concerning human variation, but it certainly helped that the defence expert was an image analyst rather than an anatomist, and that he irritated the judge by not turning his mobile phone off.

Sue testifed. The defence made their case. The girl was cross-examined. The jury deliberated and came back with a verdict that Sue had not been expecting: not guilty. Worried that she had overstepped the mark, Sue asked the prosecution barrister to check with the jury whether the science had seemed at fault to them. If it had, vein pattern analysis as a forensic technique would have to be modified or abandoned. The verdict from the jury was that it wasn't the science that was the issue. That had made sense to them. They'd gone with 'not guilty' not because they didn't believe the science but because they didn't believe the girl – she hadn't cried enough.

Instead of despairing at the fickleness of juries, Sue set about shoring up the science, so it would be better placed to combat purely emotional courtroom responses. Since CAHID were then training police officers from all over the UK in disaster victim identification, Sue decided to make the most of what she saw as a unique opportunity. She had all 500 of her police students strip down to their underwear. Her team then took photographs in infrared and visible light of feet, legs, thighs, backs, abdomens, chests, arms, forearms and hands. Once these photos had been catalogued and compared, they provided strong support for the vein pattern analysis technique.

Because police officers love trading case histories and anecdotes, news of Sue's expertise had spread far beyond Nick Marsh. Before long another Met officer turned to Sue for help in another paedophilia case. In 2009 police had searched the home of a furniture salesman from Kent called Dean Hardy. They had found sixty-three indecent photographs on his computer. Some were of South-east Asian girls, between the ages of eight and ten. All were being abused by a Western man. The metadata hidden in the photographic data files said they had been taken in 2005. The police could prove that Hardy had travelled to Thailand in 2005, and they accused him of abusing the girls. He denied it.

This time Sue Black instructed that both Hardy's hands be photographed. Then she looked at them in meticulous detail. She noted the vein patterns. She spotted a small scar at the base of one of his fingers. She looked at the pattern of the creases of his knuckles. She noted the pattern of his freckles. Then she compared her findings with the hand in the photo. They matched in every respect. The police confronted Hardy, saying, 'There is a greater similarity between your left hand and the hand in this photo than there is between your own left and right hands.' They then asked,

'Is that your hand?' Faced with such detailed evidence, this time he answered, 'Yes.'

That was the first time in UK history that freckles and veins had been used to identify a criminal. Soon afterwards documentary makers produced a programme on how to catch a paedophile, using Sue's work with the Met on cornering Dean Hardy. When the documentary was shown, four other women came forward and revealed that Hardy had abused them as children. Hardy was sentenced to six years for the Thailand abuse and at a second trial some years later, he was sentenced to a further ten years for the abuse attested to by the UK victims.

Later that year, Sue helped build the platform of evidence that convicted members of Scotland's largest known paedophile ring. Eight men from across central Scotland had been making, sharing and collecting abusive images. One of the men had 78,000 pictures on his computer. Following her work on this case, Sue and her team are currently involved in around fifteen cases a year involving the identification of paedophile predators. CAHID has become the first port of call for police who need this kind of help.

But the centre at Dundee is far from the only place where leading edge developments in forensic anthropology are being brought to bear on the identification of the unknown. At Louisiana State University, Mary Manheim is creator and director of a laboratory known as FACES (Forensic Anthropology and Computer Enhancement Services). Manheim graduated with a degree in English literature in 1981, before making a disciplinary about-turn to anthropology. She has since been involved in over 1,000 forensic cases across the US, and written three books about them, *The Bone Lady* (2000), *Trail of Bones* (2005) and *Bone Remains* (2013). She has spent decades building a database of missing people, by visiting every police department, sheriff's office and coroner's office in Louisiana. The database contains biological profiles of 600 missing people and 170 unidentified remains, and aims to make matches between the two. Now that database is connected to a nationwide resource for people looking for missing loved ones.

In one case Manheim worked on, the body of a woman was found floating in the deep water of the Gulf of Mexico, fifteen miles south of Grand Isle, Louisiana. She had been shot in the chest, wrapped in a fishing net, and sunk with a concrete anchor: it was clearly a case of murder. Though the body had been in the water some time, it was well preserved, partly because the net had kept out the crabs and fish that would normally feed on it. As Manheim noted: 'dangling body parts with mobile joints are attractive to marine life and often the first to be lost: the hands, the feet, the head'.

The body was labelled 99-15 and sent to FACES. Manheim thought it was a perfect candidate for the program and her team were quickly able to build up a picture of the woman as she would have been alive. Manheim measured her skull: her close-set eyes, overbite and oval eye sockets marked her out racially as a 'classic white European'. She was wearing an unusual turquoise and diamond necklace in the shape of a butterfly. Analysis of her skeleton revealed that she had suffered old fractures to her legs, and had an arthritic right knee which would have made her limp. She had had her wisdom teeth removed, probably by an American dentist. By measuring her leg bones and looking at her pelvis, they were able to come up with an approximate height, weight, and age. Body 99-15 had been 5'2"–5'5", aged 48–60, 125–135 pounds. The information went on to the FACES database and, in October 2004, 99-15 was identified as a 65-year-old woman who had gone missing from Missouri in January 1999. The analysis was spot-on apart from having underestimated the woman's age.

What does it feel like for a forensic anthropologist when a positive identification is made? Having spent so much time in mute communication with the dead, what is it like to share a moment with a living person who has had their worst fears confirmed? Mary Manheim knows. 'Positive identification causes family members pain, but that resolution helps them go on with their lives,' she says. Hours spent continually agonising over what their relative may be going through can begin to be spent on living their own lives.

For Sue Black, there is one identification she still yearns to make. Sue was born in Inverness, in the north of Scotland, and that's the location of a disappearance that haunts her to this day. In 1976 Renee MacRae drove away from her home in the city with her two young sons in the backseat. She dropped off the elder boy at her estranged husband's house and continued on with 3-year-old Andrew towards Kilmarnock, where she planned to visit her sister.

Neither Renee nor Andrew has ever been seen again. Later that night her empty blue BMW was spotted on fire in a lay-by on the main road south, the A9. Nothing was recovered from the burnt-out car except a rug stained with Renee's blood. Her ex-husband was interviewed and the identity of her secret lover was unveiled. An intensive search, including of more than 500 houses, garages and outbuildings in the city, revealed not a clue. Nothing seemed to take the police any closer to discovering the fate of Renee and her son.

In 2004 a television documentary aired in Scotland called *Unsolved*. It sparked a new wave of interest in the mysterious disappearance. A retired police officer came forward claiming that there had been a suggestion that the bodies of Renee and Andrew might have been dumped in a quarry close to the A9. Sue Black was involved in the operation to excavate the quarry in a painstaking search for Renee and Andrew's remains. It took three weeks to remove 20,000 tons of earth from the quarry, and cut down 2,000 trees. The operation cost more than £100,000. All that was found were rabbit bones, two crisp packets and some men's clothing.

In spite of the failure of the cold case review, Sue Black received a letter from Renee's sister, which she will always keep. 'I just want my sister home,' it read. 'I know she's dead now. I accept she's dead. Every time somebody goes looking for her my hopes are raised, and every time they don't find her, I sink deeper.' In Sue's experience, when people cannot find a family member – whether in Kosovo, Argentina, Thailand or the UK – they can never get beyond it. It's that knowledge that continues to drive her forward in her mission of bringing home the dead.

'When we come with news,' says Sue, 'it's always bad news. "It's your son"; "it's your wife"; "it's your daughter". But the bad news is tinged with a kindness that says, "Now at least you know, and you can put the body in the ground and start to grieve. And you'll never forget but you can start to move on."'

NINE

FACIAL RECONSTRUCTION

'I marvel how Nature could ever find space
For so many strange contrasts in one human face.'

William Wordsworth, 'A Character' (1800)

Never mind fingerprints or DNA. What make us recognisably individual to each other are, of course, our faces. Nature, nurture and circumstances combine uniquely in each of us to create a set of features that is the key to identification for everyone who knows us. At one time or another, we've all been misled by the similar body shape or gait or hair of a stranger; but when they turn or come close enough for us to see their face, we know our error at once. But death steals our faces from us. Our flesh decomposes, nature strips us back to the bone and the skull beneath the skin means nothing to the people who knew and loved us.

Thankfully, there is a small band of scientists whose work is dedicated to giving the dead their faces back. In the UK, Richard Neave established the technique of facial reconstruction from skeletal remains at Manchester University. He was part of a team put together in 1970 to investigate the Egyptian mummies housed at the Manchester Museum and in 1973, using plaster and clay, he rebuilt the faces of two 4,000-year-old Egyptians, Khnum-Nakht and Nekht-Ankh, known as 'The Two Brothers'. 'Right from the beginning,' wrote Neave, 'I endeavoured not to rely merely upon intuition – what was irritatingly referred to as "artistic licence".' Instead he determined the shape of the faces using average tissue thickness measurements taken from a collection of cadavers in 1898 by the Swiss anatomist Julius Kollmann.

Neave developed great skill in modelling the muscles of the face and skull, which provided a latticework for the rest of the flesh and skin to sit on. Having refined his skills in the archaeological sphere, he turned to forensic work and was involved in more than twenty cases of unidentified remains, with a 75 per cent identification success rate.

One of his most challenging cases began, paradoxically, with a headless corpse. The body of a man wearing nothing but a pair of underpants was discovered in 1993 under the railway arches of Manchester's Piccadilly Station. In spite of the best efforts of the police, his identity remained a mystery.

Three months later, a man was walking his dog across a playing field in Cannock, Staffordshire, seventy-five miles from Manchester. Suddenly the dog started digging and continued frantically until he uncovered a severed head. It had been smashed into more than a hundred pieces; later, it emerged it had been mangled by a machete. DNA tests connected it to the headless torso in Manchester, but that still didn't take the police any closer to an ID. And at first it seemed unlikely that the face could be rebuilt. A significant amount of bone was missing, especially from the crucial middle section of the skull. Police presumed the murderer had intended to make it impossible for anyone to recognise the victim of this vicious attack. But painstakingly Richard Neave glued together what remained of the skull and cast it in plaster, filling the gaps to the best of his ability and with the benefit of his extensive knowledge and experience. When the *Independent* newspaper published a photo of Neave's clay head, seventy-six families came forward thinking that they recognised the face.

The police collected photographs and details from those families and began comparing the faces of their missing relatives to the skull. As they worked their way down the list without success, it began to look as if the murderer had succeeded. Finally, they reached the last name. Adnan Al-Sane had been given such a low priority because there had been nothing about the body or skull to suggest that the victim wasn't Caucasian. But the details matched. At last, the police knew who the victim was.

Adnan Al-Sane was a 46-year-old Kuwaiti businessman who had been living in Maida Vale, west London. He came from a wealthy family and had made a fortune running a bank in his native country, before retiring at only thirty-eight. He had last been sighted, the day before the decapitated body was found, having dinner at the Britannia Hotel in Grosvenor Square, central London. Dental records and fingerprints from Al-Sane's flat confirmed his identity. The post mortem showed that he had swallowed a tooth during the attack that killed him, but his head had been hacked off after he was dead. To this day, his murder remains unsolved, the motive a mystery. But at least his family knows his fate.

Richard Neave helped to demonstrate the scientific basis for facial reconstruction, shaking off the notion that it was more an art than a rigorous scientific discipline. He spent his career working and lecturing at Manchester University, where he passed on his knowledge to the next generation, among them Caroline Wilkinson, now Professor of Craniofacial Reconstruction at the University of Dundee.

One of Caroline's own landmark cases started almost as improbably as the Al-Sane case. One August day in 2001 a sunbather came across part of a girl's body on a beach at Lake Nulde in the Netherlands. Over the next few days other body parts were found at different locations along the Dutch coast. Then a fisherman discovered a skull near a wharf, eighty miles from Nulde. The face had been mutilated beyond recognition. Investigators were baffled. They contacted Caroline, hopeful that she would agree to remake the face.

But when Dutch police told her they estimated the victim was aged between five and seven years old, she found herself uneasy about taking the case. Part of her reluctance stemmed from the fact that her own daughter was also only five at the time. But much more significant than her own emotional response was professional caution.

Back then, anatomists doubted whether it was possible to reconstruct children's faces with anything like the same accuracy as those of adults, because juvenile faces are undeveloped and lack clarity. But Wilkinson had specialised in juvenile facial reconstruction while working towards her PhD. She believed she could bring something useful to the investigation. She locked away her qualms and examined the damaged skull the Dutch police had sent over. As she studied the bones, Caroline realised the dead child had some unusual features: a large, wide nose – unlike the little upturned noses that most 5-year-olds have – and a big gap between her front teeth. Already she could see that this was a distinctive face.

In general, it's more uncommon for missing children to be recognised from photos than adults, despite the greater media coverage that they get, because their unformed faces are more similar to each other. Only one in six missing children is found because someone calls the authorities after seeing that child's picture, according to the National Center for Missing and Exploited Children, who release thousands of images of missing children every week in the US.

But Caroline was hopeful that this girl would be one of the identified ones. She applied all her skill to making a clay model of the Nulde girl's face. Photographs of the result were widely displayed in newspapers and on television around Europe. Within a week the girl had been identified as Rowena Rikkers from Dordrecht, aged five and a half.

In the wake of the identification, a horrific story emerged. During the last five months of her short and tragic life Rowena had been physically abused by her mother's boyfriend, with her mother's knowledge. She had spent the last two months of her life locked up in a dog cage. After her death, her body had been cut up and scattered throughout the Netherlands by the two people above all who should have cared for and protected her. They were eventually tracked down to Spain and convicted of their crimes. It was the first time a facial reconstruction had been used to solve a crime in the Netherlands – and without Caroline's work, Rowena's death may never have been acknowledged or avenged.

The idea of rebuilding faces is not new, nor is it always about murder. It sprung from a desire to connect with lost people by visualising them, and people have been doing it for a very long time. In 1953 archaeologist Kathleen Kenyon discovered skulls in Jericho from around 7000 BC which had clay carefully worked on to them, with shells set into their sockets to imitate eyes. She was struck by their beauty: 'Each head has a most individual character, and one cannot escape the impression that one is looking at real portraits.' The ancient Middle Eastern artists had used clay to model the physical distillation of their ancestors' identities – their faces – so that they might conquer death.

The face has always been imbued with significance. The eighteenth-century artist William Hogarth called the face 'the index of the mind'. And there's no denying that faces betray our emotions and responses – they laugh, cry, scare, soothe, entertain. The tiniest movement of our facial muscles can reveal aggression, or affection: you only have to think of the subtle difference between a confused frown and an angry frown to realise this. Our brains are highly skilled in recognising minute differences in other people's faces and we can identify hundreds of them as a result. At only five weeks old, babies can distinguish their mothers' faces. And 2.5 per cent of people grow up to be 'super recognisers', capable of identifying nearly every face they have ever seen. We can read in a face certain key elements of our humanity – gender, age, general

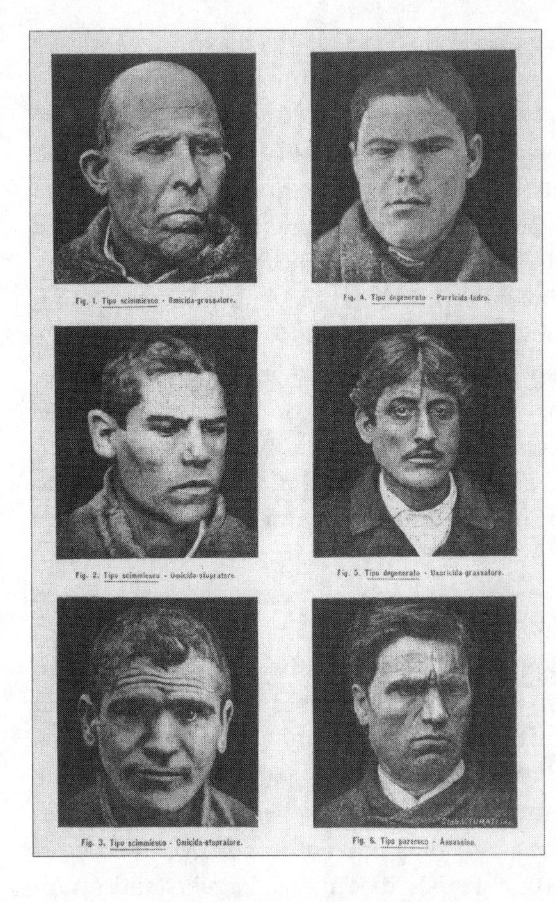

A collection of 'criminal faces' compiled by Cesare Lombroso: this plate shows murderers. Lombroso believed criminality could be predicted by the physiology of an individual

health, for example. But just because you can see someone's face, it doesn't make you a mind-reader; as Shakespeare pointed out, 'there's no art/To find the mind's construction in the face'. One thing we definitely can't tell from a face is whether someone is 'constructed' like a criminal.

The nineteenth-century criminologist Cesare Lombroso thought he knew better, however. Lombroso measured the faces of 383 lawbreakers and published a book, *L'Uomo Delinquente* (*Criminal Man*), in 1878 which ascribed to criminals 'enormous jaws, high cheekbones, prominent brow ridges, solitary lines in palms, extreme size of eye sockets and handle-shaped ears'. Later studies of Lombroso's own measurements have shown his conclusions to be nonsense. The evidence didn't back up the theory; it was based only on Lombroso's own prejudices and unfounded opinions.

But 'Lombrosia', as it came to be called, was a seductive concept, and its creator was often asked to testify at trials, with mixed success. He was outraged when one jury ignored his recommendation to convict a man of murder despite the absence of hard evidence. Although Lombroso had identified 'a physiognomy approaching the criminal type in every way', including 'outstanding ears, premature wrinkles and [a] sinister look', all of which should have been enough to convict him 'in a country less tender towards criminals', the jury was not convinced. He was also criticised by some contemporary scientists, but, despite these setbacks, his ideas were nevertheless influential. People listened to him because they instinctively seek meaning in faces.

Lombroso went about it in completely the wrong way. But, in a way, he was on the right track. In order to solve crimes and unearth the secrets of the past, scientists and investigators do need to pay very careful attention to human physiology. In Caroline Wilkinson's view, 'Any facial reconstruction produced without an understanding of facial anatomy and anthropology would be at best naive and at worst grossly inaccurate.' Painters and sculptors have long known that understanding how the facial muscles attach and move can improve the accuracy of their work, leading to a profound interest in dissection and anatomy. Leonardo da Vinci dissected thirty unrefrigerated cadavers in his lifetime, overcoming 'the fear of living in the company of these dead men, dismembered and flayed and terrible to behold'. His dissections gave rise to a series of astonishing anatomical drawings, including of a skull in cross-section, that gave Leonardo's later painterly depictions of human faces a deeper realism.

The brilliant seventeenth-century Sicilian sculptor Giulio Zumbo never saw Leonardo's unpublished skull drawings, but managed to improve the understanding of how individual faces relate to their skulls in a different way. Together with a French surgeon, he worked wax on to a real skull, leaving the 'skin' peeled back to reveal the facial muscles. The resulting full-colour model of a half-decomposed face, replete with maggots coming out of nostrils, looked uncannily like a real person.

In the nineteenth century, as we came to a greater understanding of the workings of the human body, facial reconstruction became more rigorously scientific. Early practitioners had lacked established anatomical principles to work within, so they began to generate them. German and Swiss anatomists and sculptors collaborated to interpret the relationship between the face and the skull.

In 1894 in Leipzig, archaeologists exhumed a skeleton that they thought belonged to Johann Sebastian Bach. They asked the anatomist Wilhelm His to prove it. He went about it in an original way, by taking possession of twenty-four male and four female cadavers, and laying patches of rubber at landmark points on their faces. He pushed an oiled needle through each rubber – which represented the level of the skin – and down through the face until it hit bone. Then he pulled the needle back out and measured the distance from the needle tip to the rubber. These were the world's first soft tissue thickness measurements. He averaged out these measurements and then, with the help of a sculptor, began building up clay over the skull to match them. The resulting model looked remarkably similar to contemporary representations of Bach.

Despite the scientific value of the Bach reconstruction being compromised by Wilhelm His's familiarity with contemporary portraits of the composer, his needle and rubber technique was of lasting value; the measurements he took have remained remarkably consistent, and are still used today, although facial reconstructionists think that in recent years faces in the Western world have become fatter. In 1899 Kollmann and the sculptor Büchy used the technique to rebuild the face of a Neolithic woman who had lived by a lake in Auvernier, Switzerland. The woman is regarded as the first properly scientific facial reconstruction, because Kollmann had based his model on so many soft tissue measurements, taken from forty-six male and ninety-nine female cadavers from the local area – the same tissue measurements Richard Neave would use in the 1970s to reconstruct the faces of the Two Brothers.

As the twentieth century moved forward, so did the techniques of facial reconstruction. Anthropologist Mikhail Gerasimov developed what is now known as the 'Russian Method', which pays great attention to muscle structure and less to tissue thickness measurements. He modelled muscles on to the skull one by one, and then covered them with a thin layer of clay to represent skin. He reconstructed over 200 archaeological faces – including that of Ivan the Terrible – and was involved in 150 forensic cases. In 1950 he founded the Laboratory for Plastic Reconstruction at the USSR Academy of Sciences in Moscow. It still exists, and makes an important contribution to the field.

Developments in medical technology have generated important developments in the field of facial reconstruction. X-rays and CT scans of living people have been remarkable sources of data. Until the 1980s all measurements had been taken from cadavers, which led inevitably to some inaccuracies. The walls of our cells start to break down immediately after we die, which causes fluid to drain to the back of our heads and our faces to lose their plumpness. Also, as American facial reconstructionist Betty Gatliff noted, 'When people die, they don't die sitting up, they die lying down. The soft tissue shifts.' Three-dimensional models of living faces and their skulls were always the holy grail for reconstructionists and CT scanning has provided more widely accepted thickness measurements. As a result, facial reconstruction is now more accurate – and consequently more trusted – than ever before.

Investigators call forensic artists when they've found a skull that can't be identified after crime scene clues, missing person files and forensic evidence such as DNA and dental records have led them nowhere. If investigators don't know who they are looking at, the last best hope is that a member of the public might. So it was with Rowena Rikkers and Adnan Al-Sane. A reconstructed face is a recognition tool, a memory jogger. It's not strictly 'forensic' because the reconstruction itself has no weight in a courtroom. It's only after families have contacted the police that the forensic procedure of identification begins.

But why does a face look like it does? How did it develop into this means of identification? We tend to think of the face as a social tool, which is why when we want to dismiss people disrespectfully, we tell them to 'Talk to the hand 'cos the face ain't listening' or turn our head away. In fact, our faces evolved as they did primarily for utility. Having a pair of eyes at the front of our head gives us overlapping fields of vision and thus depth perception. Our lips and jaws are perfectly evolved to chew, swallow, breathe and talk. Having one ear on either side of our head helps us to pinpoint the origin of a sound. But there are other elements, too. Familial resemblances reinforced tribal loyalties in early communities, as well as in later dynasties such as the Hapsburgs, famous for their hereditary malformation of the lower jaw.

The shape of the face depends on the twenty-two bones of the skull. The complex shape of these bones, and to a lesser extent the muscles that are attached to them, explains the variation between individual faces. Understanding the myriad variations these bones and muscles can produce is the starting point of facial reconstruction.

To deduce the shape and prominence of somebody's eyes, forensic artists look at the depth of the eye orbit and the shape of the brow. The shape of the lips and how they meet is taken from the size and position of the teeth. Ears and noses present a challenge, because cartilage decomposes after death. All we can know about the ears is where they were and whether they had earlobes; although, in life, every pair of ears is as unique as a fingerprint. It's hard to know if a nose was a button, Roman or tipped up like a pig's. But the 'bony nose' can tell anatomists a surprising amount about the 'soft nose' which sits on top of it. For example, the pointed piece of bone – the nasal spine – at the bottom of the bony nose usually has one point to it. If it has two points that makes the nose split slightly at its tip.

Facial reconstructions based on skulls have to work without the important differentiators of hair and eye colour, at least for now. Geneticists have recently learnt how to pin down nineteen different eye colours from DNA. But this information is expensive to extract – far beyond the budget allocation for reconstruction, even in a murder inquiry. DNA can also reveal hair colour but, even if that had a negligible cost, it would be of limited value to artists. Caroline Wilkinson explains: 'I took photographs of all my students this year. Only two of them have their natural hair colour. I'm forty-eight and I reckon most of my friends have no idea what my real hair colour is. I'm not sure even I know anymore.' So most artists sidestep the problem. They subtly blur out the hair (and the unpredictable ears) of their models. And yet the overall results can still be uncanny, often because of the accuracy of the soft tissue thicknesses that CT scanning has provided. The closer a model resembles a real face, the greater the chance of someone recognising their loved one. The effectiveness of an acute resemblance was proved in an extraordinary case in Edinburgh in 2013.

On 24 April Philomena Dunleavy arrived in Edinburgh from her home in Dublin. A slightly built, shy woman of sixty-six, she had come to visit her eldest son, Seamus. At his flat in Balgreen Road, they started to catch up. Seamus talked about his recent work labouring on Edinburgh's tram network. In return, Philomena tried to fill him in on the news of his four siblings. But Seamus was behaving oddly, first distracted and then agitated.

Philomena was alarmed. She told her son she was going to have a look around Edinburgh, but instead she went to Portobello Police Station. She asked an officer there where she could get a cheap room. She said, 'I don't want to spend the night with my son whilst he's having an episode.' A few days later, Seamus called his father in Dublin to say his mother was on her way home. She never arrived.

On 6 June a 24-year-old ski instructor went for a bike ride in Edinburgh's Corstorphine Hill nature reserve. The weather was hot and he decided to stop riding and find a place to sit in the sun for a while. He was pushing his bike along a narrow path when he saw a set of brilliant white teeth gleaming at him from the dirt. The teeth were set in the remains of a severed head. Most of the flesh had rotted away, but the carrion-loving flies were still there.

In the shallow grave that the shining teeth had revealed, forensic anthropologist Jennifer Miller unearthed two severed legs and a human trunk, which she ascribed to a woman of about sixty. She noted that the brilliant teeth were the result of expensive cosmetic dentistry. One of the rings she removed from the cadaver was a traditional Irish Claddagh ring. Equipped with this limited information, the police spent weeks searching missing persons lists.

Finally they asked Caroline Wilkinson to make a facial reconstruction, which she did using 3D scans of the skull, then filling in the soft tissue digitally. The resulting image was circulated to police forces throughout Europe, and shown on BBC *Crimewatch.* The presenter of *Crimewatch* also mentioned the Claddagh ring, which made a family member in Dublin feel doubly sure that she was looking at Philomena. The likeness of Wilkinson's image was uncannily accurate. The identity of the body was put beyond doubt by dental records.

A few days later Seamus was arrested and charged with his mother's murder, which he denied.

The jury didn't believe him. Instead, they accepted the prosecution's case that Philomena had gone back to Seamus's flat sometime after she'd spoken to the police. There, she had died. The pathologist noted damage to the small bones in her neck (which often signifies strangulation), injuries to her head and smashed ribs. Seamus had cut off her head and legs with a saw. But it was impossible to tell whether these injuries had been sustained before or after death. A journalist for *Herald Scotland* reported on the more disturbing possibility: 'Philomena Dunleavy may still have been alive, but unconscious, when her son began to hack off her legs.' The exact circumstances of her death will never be known.

What we do know is that Seamus then put his dismembered mother's remains in a suitcase, and took her up Corstorphine Hill. He hacked out a shallow grave with a spade, and dumped his mother in it. As forensic experts often tell us, murder is easy compared with the difficulty of disposing of a body effectively. It was only two months later when her body resurfaced and, with it, the vital clues that would lead to his conviction. The prosecutor called it 'a case in which pieces of evidence came together like strands in a cable'. In January 2014, Seamus Dunleavy was convicted of murder, in no small part because of Caroline Wilkinson's work.

Such a swift identification of a victim is not guaranteed. On 18 November 1987, a cigarette butt ignited some rubbish under a wooden escalator at London's busiest train station, King's Cross. The fire grew in intensity until the escalator vented a fireball burning at 600°C, which hurtled up the escalator and into the underground ticket hall above it.

Hundreds of people were trapped in the complex of tunnels connecting King's Cross's six Tube lines. Some took the escalator up to escape the black smoke underground, and were burned alive. Others thumped on doors to try and get on trains that didn't stop. When firefighters finally fought back the blaze, they discovered thirty-one dead bodies.

Over the following days and weeks the police managed to identify thirty of the dead. But one middle-aged man eluded them. Richard Neave was asked to reconstruct the man's face, which had been terribly burnt by the fireball. He found some pieces of tissue around the nose and mouth which helped him predict the shape of that part of the face. And he was given an extensive dossier outlining the victim's height, age and state of health.

Interpol were approached for help and enquiries were made as far away as China and Australia. Richard Neave's reconstruction was shown in all the major newspapers in the UK and hundreds of people phoned up, believing it to be someone missing from their circle. But no definite matches were possible. Meanwhile the body was buried in a grave in north London marked 'AN UNKNOWN MAN'.

In 1997 Mary Leishman, a middle-aged Scot, made enquiries about her missing father, Alexander Fallon. When his wife had died in 1974, Fallon's life had fallen apart. He had been unable to cope with everyday life. He had sold his house and ended up sleeping rough on the streets of London, among thousands of other virtually anonymous homeless people. Mary and her sister had begun to wonder whether the unknown victim of the King's Cross fire might be her father, but she wasn't hopeful. At the time of the fire her father had been seventy-three years old and five foot six, whereas the post mortem had put the dead man at between forty and sixty years old and five foot two. Yet the corpse had smoked heavily, as had Alexander Fallon and, like him, it had a metal clip inside the skull as a consequence of brain surgery. At the time of Mary Leishman's enquiry the police thought they had a match with another missing man, Hubert Rose, so they didn't follow up her query. Then, in 2002, a service commemorating the fifteenth anniversary of the victims of the fire was held in north London. This nudged Mary Leishman to raise her concerns with the police again.

In 2004 Richard Neave was shown photographs of Mary Leishman's father. He rifled back through his records to find photographs of the mystery victim's skull, and his own clay model. He compared frontal and profile photos and saw the similarities immediately – both had prominent cheek bones, thin lips, similar spacing between the eyes, the same laughter lines running from the corner of the mouth to the chin, although the man in

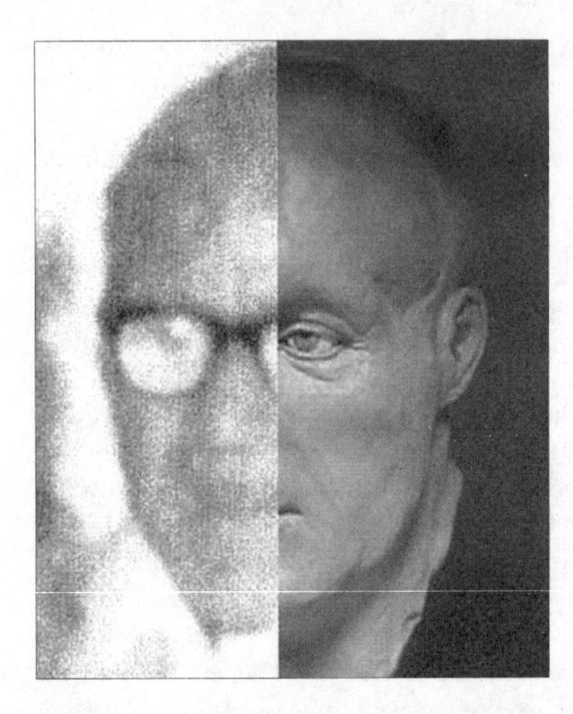

A photograph of Alexander Fallon, a victim of the King's Cross fire, compared with the facial reconstruction created using his remains

the photograph had a much more bulbous nose than his model. With the further corroboration of dental records and of the neurosurgeon who had inserted the metal clip, the final victim of the King's Cross disaster was finally identified as Alexander Fallon – sixteen years after his death.

Richard Neave's model of Alexander Fallon prompted his daughter Mary's enquiries. That was all it was designed to do. A string of other factors, including documentary evidence, supported the identification and made a traumatic exhumation unnecessary. And, as Mary Leishman said: 'One thing that makes us feel certain now that my father was a victim of the fire is that we have, with the help of the police, established that no benefits were uplifted in his name after the date of the fire. If my father was alive, he would have been first in the queue whenever there was money to be had.'

◆ ◆ ◆

If the King's Cross fire happened today, Alexander Fallon's face would be rebuilt by computer. Digital modelling has not replaced clay modelling – which Caroline Wilkinson still teaches to her students in Dundee – but nowadays 80 per cent of forensic facial reconstruction is computer-based.

First, Caroline will scan the skull in three dimensions, usually with a CT scanner, and then import the resulting model to an image-editing program. Then she chooses one of a number of basic muscle templates and overlays it on to the skull. Now Caroline tweaks the muscles manually – click, drag, click, drag – based on the same standard thicknesses that she uses when working in clay. Computer modelling is quicker than clay modelling because having the templates means Caroline doesn't have to start from scratch every time. But not by much. It takes a long time to add skin, eyes and hair, and to texture them properly.

But there are advantages to the computer method other than speed. Caroline can vary elements such as skin tone and hair colour, then print out a dozen possible images for investigators to look at. Three-dimensional scanning allows the reconstructor to see injuries to the skull, such as a hammer mark, more clearly than plaster casting. With precise modelling of the wound and the weapon it's possible to make a model of the event as well as the face, which can be shown in a courtroom further down the line. If someone recognises a reconstruction and sends in a photo of their missing loved one, artists can scan it and superimpose it on to the skull. This is the digital version of the technique that was used for the first time to incriminate Doctor Buck Ruxton in the Jigsaw Murders of 1935 (see pp. 48–51).

Craniofacial modellers don't only use computers to create a face as it once was, but also as it might now be, especially in the case of missing persons. The process of 'age progression' can be automated to a significant degree. Our ears grow longer as we grow older, at a more or less predictable rate, and algorithms exist to plot the basic sagging and puffing of an aging face. But age-progressed images are largely down to the instincts and experience of the artist, who looks at sequences of photographs of

The 'Butcher of Bosnia'. From left: former Bosnian Serb leader Radovan Karadžić in 1994; as he appeared while evading capture after his indictment for war crimes; and at the International Criminal Tribunal for the Former Yugoslavia in the Hague in July 2008. He was charged with 11 charges of genocide, war crimes and crimes against humanity

people as they get older and identifies general trends. The artist uses photos of older siblings for guidance, adapts the image to reflect the kind of life a subject may have led, and adds distinctive clothing or facial hair. Fine details like liver spots can be added manually, too. For Caroline Wilkinson, 'The most difficult things to work out are skin colour, eye colour, how fat or thin they are, and whether they have wrinkles.'

The hunt for missing people can also be hampered by changes to their appearance that have nothing to do with aging, and which can be effected by techniques as simple as growing facial hair. Radovan Karadžić is a former Bosnian Serb politician who was indicted for war crimes in 1995 by the International Criminal Tribunal for the Former Yugoslavia. Among other atrocities, Karadžić was charged with ordering the 1995 Srebrenica Massacre in which 8,000 Bosnians were murdered. After his indictment, the 'Butcher of Bosnia' disappeared, shaved his hair, grew a beard, donned a priest's robe and lived an itinerant life, wandering from monastery to monastery.

Caroline Wilkinson was asked to make an age-progressed image of Karadžić. She got the shape of his face spot on, but underestimated his beard. He had moved to Belgrade and started wearing his long hair in a ponytail, donning large square glasses and hiding behind an enormous white beard. Calling himself 'Dabić the Spiritual Explorer', he masqueraded as an expert in human quantum energy, worked in an alternative medicine clinic and gave public lectures. But the age-progressed images gave a new impetus to the hunt for Karadžić. In 2008, a year after Caroline had sent them her image, he was arrested by Serbian security forces, and extradited to The Hague to stand trial. The trial is still going on.

Computers often help forensic artists identify less heinous criminals, too. They analyse CCTV footage and compare the images to a suspect. When offenders do not simply break down and confess at the sight of their blurred selves on video – which they often do – it is difficult to prove conclusively that it was them. Even when footage is high quality, identifying an unfamiliar face by sight is not the most reliable of procedures. Computerised facial image comparison can offer a more reliable alternative. One method is to superimpose a still from the video on to a photo of the suspect, although that can be awkward when criminals haven't been looking straight at the camera, which they tend not to do. Another technique, which has been used for approximately the last fifteen years, is called photoanthropometry. This involves comparing the proportional distances and angles between landmarks on two facial images. But the technique is not perfect. Even when the suspect is asked to pose for their photo in the same alignment as the person in the video, there is a complex bundle of variables to accommodate, such as distance from the camera, camera angle and head orientation.

We have seen how forensic artists identify the dead from their skulls, missing people from photographs, and wanted people from video footage. One other significant aspect of their work is to depict wanted people based on eyewitness accounts. Historically, this was the job of a sketch artist who would translate an often shaky witness recollection into a drawing of a suspect. But in the 1980s researchers at the University of Kent helped develop an alternative method known as E-FIT (Electronic Facial Identification Technique). Police forces around the world now use E-FIT and it makes regular appearances in the media. To make an E-FIT, an eyewitness looks at a swatch of computer-generated faces and clicks on the one that looks most like the person they saw. They are then presented with another, more narrowly defined set of faces. In this way the image is refined, until it is a relatively close representation of the person the witness remembers.

Facial reconstruction began as a way of bringing us face to face with our history – and we are still using it for that purpose. In 2012, a set of bones was found underneath a car park in Leicester. They were suspected to belong to Richard III, the last Plantagenet king of England, who died at the nearby Battle of Bosworth Field in 1485 and had been buried in a local church.

The Richard III Society assembled a team to investigate the remains. Scientists began analysing DNA samples, and scanning the skull in three dimensions. They sent the digital skull to Caroline Wilkinson, who set to work on making the king's face, avoiding looking at existing portraits of him so as not to contaminate the scientific process. She and her team modelled the muscles and skin using stereolithography, a computer process whereby a moving laser beam builds up a structure, layer by layer, from a liquid polymer that hardens on contact with laser light.

When the DNA results came back, and matched with a descendant of the king, Caroline finally compared her model to portraits. It was strikingly similar, with its arched nose and prominent chin. 'It doesn't look like the face of a tyrant,' said Philippa Langley of the Richard III Society. 'I'm sorry but it doesn't. He's very handsome. It's like you could just talk to him, have a conversation with him right now.'

Caroline is proud of her work on Richard III. 'Our facial reconstruction methods have been blind tested many times using living subjects and we know that approximately 70 per cent of the facial surface has less than 2 mm of error,' she reveals. To reach that level of accuracy Caroline stands on the shoulders of all the face makers that have come before her, from Giulio Zumbo to Wilhelm His and Richard Neave. But it's her own artistic obsession with observation that has helped her to do it so effectively. She describes herself as 'really annoying to go out with because if I'm watching a film I spend all my time going, "Oh, look at his ears, look at his nose, what a great nose," and everyone's going, "Shut up! Just watch the film." When I'm on the train I quite often get my phone out and take sneaky pictures. I get my iPad out, pretend I'm reading something and take a picture, I'm terrible.

'I also collect photographic portrait books wherever I travel overseas, which I do mainly for my archaeological work. The places I visit have books of photos that you can't get on the internet. So if I go to Egypt I'll try to buy a book of pictures of Egyptian faces, and so on. So now I've got this great database of faces that we can use to inform what we do.'

And it's that access to such a vast range of faces from all over the world that makes our present-day forensic artists more useful artist-anatomists than Leonardo da Vinci could ever be. It's the application of science to the world of artistic representation that makes it possible for the dead to tell yet another chapter of their story to us.

TEN

DIGITAL FORENSICS

'The advent of the internet has complicated mystery plotting because so much more information is available to both the sleuth and the reader. The reader is not likely to remain long interested in a detective who is too dumb to take the obvious first step in any research – going on line for related information.'

Jeffrey Barlow, Berglund Center for Internet Studies

Angus Marshall and his wife are forensic scientists. People at dinner parties assume they spend all day in the morgue dissecting bodies. Shirley Marshall soon disappoints them when she explains that her DNA work is almost entirely lab-based. Angus lets them down still further: 'The only times I cut up anything fleshy is making dinner, or working on my car, and the second one's an accident.'

At school Angus joined the radio club to get his hands on electronics. One day a maths teacher brought a microcomputer in to show the class. 'It led to the establishment of the computer club and that's been my downfall. I haven't seen daylight properly since about 1983.'

After graduating, Angus began work as a computer scientist. At the University of Hull he was stationed at the Centre for

Internet Computing, whose name made it simply irresistible to hackers. One even managed to wipe out the internet connection for the whole of the university's main campus. Angus set about tracking the hacker, and managed to trace his IP address all the way back to his street address in Amsterdam. These seem humble enough beginnings, but Angus was proud of the results of his dogged research and submitted a report of the investigation to the British Forensic Science Society. So when a far more serious and disturbing case arose, they knew who to call.

Thirty-one-year-old Jane Longhurst lived in Brighton, where she worked as a special needs teacher. She wore her chestnut hair neatly down to her shoulders. Everyone knew her as gentle and bubbly, especially her friends in the local orchestra, where she played the viola. Early on the morning of Friday 14 March 2003, Jane kissed her boyfriend Malcolm goodbye as usual.

When he came back that evening to find her gone, Malcolm quickly became concerned. Jane was dependable. She let people know what her plans were so they wouldn't worry. At midnight he was so disturbed by her absence that he phoned 999. The police initially treated Jane's disappearance as an ordinary missing person case, but after five days they changed it to a major murder investigation. Jane's bank said that none of her accounts had been touched since Friday. And her network provider could tell that her phone was turned off because it hadn't once communicated with any of their transmitters.

After a month of searching, involving seventy police officers and numerous newspaper appeals, Jane's body was found on 19 April. She had been dumped in a wooded nature reserve in West Sussex, and set ablaze. A passer-by who saw the flames had called the fire brigade. The fireman who found the body noticed a pair of nylon tights dug deep into Jane's neck. CSIs scouring the area found a match and an empty petrol canister.

Jane had to be identified from her dental records. When they examined her body, the two pathologists noted that the tights were pulled so tightly around Jane's neck that they had broken the skin and caused bleeding. A few days later, the police arrested

Graham Coutts, a door-to-door salesman of cleaning products, and charged him with Jane's murder. He was the guitar-playing boyfriend of Jane's best friend, and had known Jane for five years.

When Coutts was confronted with the pathologists' reports and the trace evidence, he said nothing at first. But eventually he admitted to killing Jane. He had arranged to take her swimming at the local leisure centre, he told police, but instead took her back to his flat for a cup of tea. There he wrapped the tights around Jane's neck in a consensual act of erotic asphyxia, gradually tightening the ligature as he masturbated. Once he had reached orgasm he looked at her body and noticed 'to my horror' that it was lifeless. He then put her body into a cardboard box, and moved it into his garden shed.

Eleven days after Jane's disappearance, the police visited Coutts. They were trying to question everyone she knew, hunting for clues. He decided at that point that he had to move her body to a room at a nearby Big Yellow Storage facility, for which he paid for an 'all hours' key. Over the next three weeks he visited Jane's body nine times. When the stench of decomposition grew too strong, he moved her again, on 17 April, to the nature reserve, where he set fire to her remains.

When they investigated the storage unit the police found Jane's mobile phone, purse, jacket and swimming costume, and a shirt belonging to Coutts with her blood on it. They also found a condom containing his semen, and with her DNA on the outside. They searched his flat and took away two computers. Together with the Police Computer Crime Unit, Angus Marshall went to work on them, fighting against his own emotional reaction to the hideous things Coutts was charged with.

In court the defence argued that Coutts was guilty only of manslaughter, and called forensic pathologist Dick Shepherd (see pp. 77–86) to the stand. He testified that in acts of erotic asphyxia it is possible for someone to die quickly, within a second or two, as a result of the inhibition of the vagus cranial nerve. The pathologist for the prosecution, Vesna Djurovic, denied this possibility, arguing instead that it takes two to three minutes for a person to die of strangulation – plenty of time for Coutts to know exactly what he was doing.

One of Coutts's ex-girlfriends testified that he had partially strangled her on many occasions during their five-year relationship. Two of Jane's ex-boyfriends gave accounts of ordinary sex lives with her. When cross-examined by the prosecutor, Coutts admitted that he had a fetish for women's necks, and that this was the first time he and Jane had engaged in a sexual act.

For Angus the case was proving very difficult, both emotionally and professionally. He was propelled 'from a relatively trivial hacking incident into a very nasty murder. I will never forget that case.' It was career-changing. It also gave him a chance to see the kinds of things people thought they could do with impunity, and then to try and unpick that impunity. There were lessons aplenty for him: 'I was cross-examined by the two barristers. They were having problems with the concepts and weren't asking the questions in the right way. So the judge stepped in because he understood the technical issues far better than they did.'

Unfortunately, the judge asked Angus one question about the use of cookies – the little bits of data stored on your computer that communicate with the websites you revisit. It spooked the jury. 'They started passing notes to the judge, wanting to know how they could protect themselves and hide their online activities from spouses and other family members.' Once the judge had restored order, Angus proceeded to give his evidence.

He had found more than 800 pornographic images on Coutts's two computers, of which 699 were of strangled, suffocated or hanged women. One showed a Father Christmas strangling a girl. As well as finding the images, Angus had pieced together a timeline of Coutts's online activity. He had been visiting violent pornography websites, such as 'Necrobabes', 'Deathbyasphyxia' and 'Hangingbitches'. The frequency of his visits had increased in the weeks before Jane's death, when he also paid for memberships of websites such as 'Club Dead' and 'Brutal Love'. His visits and downloads reached their peak during the day before Jane's death, and the two days before her body was found in flames.

Graham Coutts was convicted of murder and sentenced to life in prison. Angus recalls the judge commenting on the importance of 'the evidence from his computer showing his normal patterns of activity and the total disappearance of that pattern on the day of the murder'. Since the case, Angus has made pulling together timeline evidence a priority.

Violent criminals often leave digital traces of the twisted paths down which their minds go. Does the internet spur them along those paths? At any one time there are roughly 100,000 'snuff' sites on the web, disseminating images and video of killings, cannibalism, necrophilia and rape. The UK and US governments have taken steps towards combating these kinds of sites – though both more tentatively than Iceland's, which has attempted a complete ban on online porn.

However vigilant the authorities are, the problem remains that when a site is shut down it usually opens up again under a different domain name almost immediately. Getting to the root of the issue, and going after the producers of violent pornography, needs a level of organisation and international co-operation which has so far not been met. There are people who argue that sites peddling violent pornography only exist because there is an appetite for them. The relationship between the sites and the appetite still needs research and clarification, but to call it anything less than reciprocal seems deluded. Whether these internet images cause extreme behaviour or merely mirror what already exists, there is no doubt that violent sex offenders use them to fan the flames of their own fantasies.

On the evening of 26 May 2013, 23-year-old Jamie Reynolds sent a short text message: 'I'm excited. Don't be late.' He had asked 17-year-old Georgia Williams, the daughter of a police detective, to come over to his home in Wellington, Shropshire, to model some clothes for a photography project. Reynolds didn't tell Georgia that he'd been planning the project for months.

When she arrived he gave her high heels, a leather jacket and leather shorts to put on. He took some photos and asked her to stand on a red recycling box on the landing. Round her neck he placed a noose, which he attached to the loft hatch above. He took a photo. At this point, according to police who saw it later, Georgia looked 'happy' and 'compliant'. Then Reynolds kicked the box from beneath her. A bruise found in the small of her back suggested to a pathologist that he had applied downward pressure with his knee to speed up her suffocation. He then sexually violated her body.

When police examined Reynolds' computer they found dozens of composite images. He had taken the heads of innocent girls on Facebook and put them on to bodies engaged in hardcore pornography. They found 72 violently pornographic videos, almost 17,000 images and 40 fantasy stories written by Reynolds, one of which was called 'Georgia Williams in Surprise'. Reynolds had taken photos of his victim before, during and after the attack. The prosecuting lawyer asked for the material not to be shown in open court and only to be viewed by the judge because of its severely distressing nature. Reynolds was given a life sentence for an act which Georgia's father called 'horrific and beyond comprehension'.

The global expansion of personal computer and smartphone ownership has made it much easier for people like Graham Coutts and Jamie Reynolds to indulge their perverted fantasies. But the majority of people use the internet to do relatively innocuous things (even if the jury's reaction to Angus's explanation of cookies at the Coutts trial might suggest otherwise). Criminals, too, use the internet to do ordinary things. They write emails to family and shop with online retailers. But when they step down illegal paths, they leave a footprint that forensic digital analysts like Angus can decipher more clearly than many of them realise.

Today's torrent of personal devices began as a trickle. In the early 1980s forensic digital analysts mostly helped police investigate copyright infringement – such as kids copying games for their Atari game consoles – and fraudulent business activities.

Hard drive storage capacities were so small back then that an expert could often browse through all the files on a drive until he found what he needed to secure a conviction. 'Computers were relatively dumb devices at first,' says Angus. 'The complexities, the interactions that we see today, just weren't there.'

Up until the mid 1990s computers connected using the dial-up 'bulletin board system', which was a precursor to the world wide web. People used bulletin boards to talk to other geeks about technical problems they were having, or to get help with completing a game they were playing. There were a few renegades exploring the possibilities of using their new-found powers for evil, but most people were just excited by the possibilities. To be involved, you needed a fair amount of technical expertise and you often had to build quite a lot of the kit yourself.

But computing power continued to grow exponentially. When Microsoft launched Windows 95, they opened up the world wide web to ordinary people. At this point the police started to take digital forensics seriously, realising, like Angus, that 'criminals tend to be very good at adopting new technology'. In the early 2000s the Home Office and FBI were both devoting substantial resources to establishing units capable of tackling cyber-crime. Home Secretary Jack Straw said at the launch of Britain's National Hi-Tech Crime Unit, 'New technologies bring enormous benefits to the legitimate user, but also offer opportunities for criminals, from those involved in financial fraud to paedophiles.' The National Hi-Tech Crime Unit took on new crimes that the digital revolution had made possible, like hacking, and old crimes which it had facilitated, like stalking.

In 2006, the national unit was replaced by regional ones. Today at a crime scene the lead detective decides whether she needs someone from her force's hi-tech crime unit to look at the digital material. 'Just like with DNA,' Angus explains, 'when they've got eyewitness accounts, fingerprints and everything else, they often don't need the expensive analysis. But for something like stalking or grooming, they have to go to hi-tech crime.' When a unit doesn't have the capacity or expertise to analyse digital evidence, the detective will call in an independent like Angus. By that point 'the routine work has often been done. Most of the time investigators want immediate answers to difficult problems, so I improvise and invent new techniques as I go along.'

An example of this improvisational approach happened in a recent child abuse trial. The accused man – let's call him David – was charged with multiple counts of paedophilia. His defence strategy was to discredit the key witness, his step-daughter 'Sarah'. He claimed it wasn't he who'd had sex with the 14-year-old, it was the boys she'd been having dirty chats with on Facebook. As evidence in support of his claim, David produced data from a 'key logger' which he'd installed on Sarah's computer. A key logger is a hidden program which silently records the actions of the computer user. Every time Sarah typed something or clicked on something in her web browser, the key logger would capture a screenshot – a complete picture of everything on the screen. David had periodically downloaded these screenshots. Some of the ones he presented to the court showed an indecent Facebook chat session between Sarah and a teenaged friend of hers, 'Fred'. But both teens vehemently denied that the chat had ever taken place.

Angus more often examines the digital life of suspects than of alleged victims. But in this case the best way to corroborate or invalidate David's testimony was to look at Sarah's computer. On it he found no evidence of a chat with Fred, but that didn't mean it hadn't taken place. 'As a general rule these days Facebook doesn't leave traces behind on hard drives. Everything happens in the browser,' Angus explains. While he did find a key logger installed on the computer, the screenshots of the alleged chat were not present. But that wasn't evidence one way or another either, because key loggers usually delete screenshots once they've collected a certain number of them to prevent the hard drive from clogging up.

However, Facebook itself keeps records of all chats even if users have deleted them, and Angus considered asking the company for the chat histories of Sarah and Fred. But that would have fallen very close to communications interception and covert surveillance, so he would have needed authority under the Regulation of Investigatory Powers Act (2000). Then Facebook Inc. would doubtless have taken their time. Angus would have had to wait six months or more for what he needed.

Next he asked Sarah for her login details, signed into her account and found no trace of a conversation with Fred. Of course, it might have been that she'd deleted the conversation. But what she could not do was completely expunge anyone from her lists of 'friends'. On Sarah's 'current friends', 'deleted friends' and 'requested friends', Fred was nowhere to be seen. Using Fred's login details, Angus found no trace of a conversation or friendship with Sarah on his account either. On Sarah's account Angus did, however, find records of other, milder conversations with other boys that David had provided screenshots of. It seemed as though David had nestled fabricated screenshots among real ones. But Angus was all too familiar with the principle that absence of evidence doesn't mean evidence of absence.

In the end, Angus wrote to the judge reporting that he couldn't be certain what had happened. It was theoretically possible that Sarah and Fred had had the indecent chat under false profiles that looked identical to their normal ones. Equally, David, who was a good amateur photographer, could have forged the screenshots. To gain a satisfactory view of what had happened, Angus needed to look at David's computers to see if he had manipulated the screenshots with a graphics editing program.

At this point the judge had to make the call. Should he continue with the trial? Or should he suspend the sitting and keep the jury sequestered for another week while Angus examined David's computers? He decided to proceed. The jury listened to the remaining testimony of the victims, and to Angus's evidence. Whilst his evidence was inconclusive – and he was careful to make that clear to the jury – it formed another piece of possible evidence that David was a manipulative liar. The jury deliberated, and found him guilty. He is currently serving twenty years in jail.

◆ ◆ ◆

As the case of the key logger shows, the more people who use the increasing number of functions available on their digital devices, the harder it becomes for forensic digital analysts to do their job. Whereas some forensic scientists are able to answer straight questions – 'Does this blood belong to Mr A or Mr B?' – people in Angus's area of specialization have to judge the authenticity of evidence, construct timelines of online and offline activity and assess the validity of alibis. Those without the right blend of imagination and vigilance need not apply.

Angus loves the job for its intellectual challenge. 'I'm always learning something new, not just grinding away doing the same thing day in day out, but solving problems.' The hardest thing for him to bear is when his investigations throw up nothing. 'I don't know of anyone in the business who, when faced with a no result job, will stop. You keep probing and probing and probing because there must be something there, there's always something there, and it's really hard to accept that you've done everything you can and hit the limit.'

Before Angus can go to work, he needs something to work on, and getting it can be a headache. 'In order to collect evidence against one bad apple, you cannot storm in and seize the computer of every employee in an office. The response has to be proportionate.' Laying hands on the hardware for Angus to work on is the job of the police. They have to justify a search warrant so they can confiscate digital devices from the suspect's living room, or trouser pocket.

When a device is found at a crime scene it is often covered in fingerprints and DNA. But because the magnetic brushes that CSIs use to powder up and expose fingerprints emit electromagnetic fields, they can destroy evidence within the device. Hence, CSIs have learned to place devices carefully in antistatic plastic bags, then send them to the digital analysts. 'We still occasionally encounter devices sent to the wrong unit,' says Angus. 'I've seen mobile phones sent to the CCTV unit because detectives wanted the photographs. I've seen officers pick up a mobile phone – very very rarely now, but I have seen it – and start poking at it themselves to see what's on there.'

Once an uncontaminated device has found its way to the hi-tech crime unit, then, according to Angus, 'unless it's a really high priority job like a murder or live missing persons case, it will sit in a storeroom for about six months, because forces have so much work to do'. But technology is constantly evolving, and now there are software programs that can provide a forensic examination result almost instantly. The device that makes its way to Angus nowadays is seldom an answering machine, printer or fax machine. Usually it's a computer, smartphone or tablet. These tiny devices contain a detailed (if partial) snapshot of a person's life. To damage them can be to damage justice. 'Rule One is always, as far as possible, *preserve*,' Angus notes. As well as for forensic digital analysts, this is the golden rule for CSIs and civilians who want to provide admissible evidence. In practice, this usually means forensic analysts will make a direct copy of the contents of a machine they are going to investigate, in order to preserve the integrity of the original.

When the term 'forensic computing' was first used in 1992, it was in relation to recovering data from computers for use in criminal investigations. In one of Angus's early cases, a company director had accused previous directors of fraud, and collected the company's main hard drive to present as evidence. He had sent the drive for a 2-week repair, stored it at home for a week, then finally given it to a forensic computing firm for examination. Angus reported to the judge that this chain of evidence preservation was not good enough. It was impossible to be sure that the employee hadn't added, altered or overwritten files at some point in the drive's complicated journey. As Angus neared York Station on the train down to Leeds Crown Court for the hearing, he received a phone call telling him that the judge agreed with his report and had dismissed the case. He got off at York, walked across to the opposite platform and headed back home to Darlington.

'Sometimes I have to break Rule One,' says Angus. 'The latest iPhones and BlackBerrys are virtually impossible to copy. I have to install software on them to "jailbreak" them. Then Rule Two comes in: If you can't copy it and you're going to have to alter it, make sure you know what you are doing and can explain it. Contemporaneous notes is the charm.' If a careless investigator opens a file, the time is recorded on the file itself. This hinders the creation of timelines and, as adversarial lawyers love mentioning in court, fundamentally alters the file.

Once Angus has an immaculate copy of the hard drive, he uses specially tailored software to look at both the current files and deleted files. From computer and smartphone drives Angus can restore almost all deleted photos, videos and messages, just as an old-school detective might have retraced the impression of a rubbed-out pencil line on a letter.

On mobile phones Angus will look at text messages, called numbers and missed calls. Text message dialogues sometimes show what criminals were saying to each other around the time a crime was committed. Individual text messages can provide crucial evidence, too. On the morning of 18 June 2001, 15-year-old Danielle Jones went missing near her home in East Tilbury, Essex. Suspicions quickly fell on her uncle, Stuart Campbell, and he was arrested when investigators found a green canvas bag in his loft containing a pair of white stockings tainted with a mixture of both his and Danielle's blood.

Campbell claimed that he had been at a DIY shop in Rayleigh, a half-hour drive away, when Danielle went missing. Police examined his mobile phone and found a text message sent from Danielle's phone that morning:

> HI STU THANKZ 4
> BEIN SO NICE UR THE
> BEST UNCLE EVER!
> TELL MUM I'M SO
> SORRY LUVYA LOADZ
> DAN XXX

But when police interrogated the records from the network providers, they found that both his and Danielle's phones had been within the narrowly defined range of the same mobile phone transmitter when Campbell's phone received the text message.

Linguistics expert Malcolm Coulthard demonstrated in court that Danielle habitually wrote her text messages in lower case. He also noticed that in another text on Campbell's phone, sent shortly after the first, the word 'what' had been shortened to 'wot', whereas Danielle always typed 'wat'. Clearly, the text message had been planted and Campbell's fabricated evidence had imploded. Despite the fact that a £1.7 million search operation by Essex Police failed to discover Danielle's body, her uncle is now serving life behind bars.

Accurately locating victims and suspects at the time of a crime has obvious benefits for investigators. Modern iPhones and Android phones log their movements by default, making it possible to plot a detailed map of where somebody's phone has been – and, by assumption, where they have been, too. The location-tracking feature can be disabled deep in the smartphone's settings, but many people don't know this. The iPhone 5S has a specialised location chip that runs off reserve battery power. Users have reported their iPhone continuing to track their movements for four days after the phone has run out of battery and turned itself off. The justification for the location data is that it helps Apple to improve its maps app, and to tailor suggestions for things for users to do nearby. Needless to say, the police are interested in this data too.

Even if a user turns off location tracking on their phone, investigators can interrogate network provider records to fix an approximate area at a given time. This is because mobile phones constantly communicate with local phone towers in order to find a signal. These towers tend to cover small areas, as occurred with Stuart Campbell in East Tilbury – and also in a remarkable case in Scotland in 2010.

On the morning of 4 May, 38-year-old Suzanne Pilley set off on her way to her job as a bookkeeper for a financial services company on Thistle Street in central Edinburgh. At 8.51 a.m. she was caught on CCTV coming out of Sainsbury's, where she'd bought her lunch. And that was the last time anyone saw her alive. Anyone, that is, apart from her work colleague 49-year-old David Gilroy. Gilroy was married with children and had been having an affair with Suzanne for about a year. She had recently decided to end their relationship for good, having had enough of Gilroy's controlling nature and fits of jealousy.

In the month leading up to Suzanne's disappearance, Gilroy had bombarded her with more than 400 texts and numerous voicemail messages. He had been desperate to keep the affair going, and unwilling to accept her rejection. On two particular days he had sent more than fifty pleading texts. The day before she vanished, Gilroy had left her numerous texts and a voicemail message in which he said, 'I'm worried about you.'

Suzanne had spent the night before her disappearance with a new man, Mark Brooks, which sent Gilroy over the edge. He murdered Suzanne in the basement of their office, and hid her body in the stairwell. He made an excuse to his colleagues – who later described him as 'seeming clammy, with scratches on his neck and face' – to take the bus home and collect his car. On his way, CCTV footage showed him buying four air fresheners from Superdrug. Back at the office, Gilroy altered his engagements so that the next day he would have to drive 130 miles into the rural heart of Argyll to check on a school whose accounts his firm was keeping. Then he bundled Suzanne's body into the boot of his car.

That evening he went to see one of his children perform in a school concert, then on to a restaurant with his family. Meanwhile, Suzanne's worried parents had reported her missing.

On 6 May, the police interviewed Gilroy. They noticed a cut on his forehead, subtle bruising on his chest and curved scratches on his hands, wrists and forearms. Gilroy said he had scratched himself while gardening. Forensic pathologist Nathaniel Cary would later examine photographs of these injuries and testify that they could have been made by another person's fingernails, possibly in a struggle, and that he had seen similar scratches on stranglers before. He added that he couldn't be sure because Gilroy had covered the scratches in flesh-coloured make-up. But he did concede under cross-examination that Gilroy's version of how he got the scratches was possible.

At the time, the police were suspicious enough to seize Gilroy's mobile phone and car. When forensic scientist Kirsty McTurk opened the car boot, she noticed a fresh smell coming from it, like 'air freshener' or a 'cleansing agent'. She looked for evidence in the boot and then in the basement stairwell at the office in Thistle Street. She could find no trace of Suzanne's DNA. However, when specially trained cadaver dogs smelled the boot and the stairwell they showed 'positive indications' of detecting human remains or blood. One of the dogs, a Springer Spaniel named Buster, had previously managed to locate a dead body in nearly 3 metres of water.

Police also found vegetation and damaged suspension underneath Gilroy's car. The roadside cameras were inconclusive, but detectives felt certain he had made a detour off the A83 Rest and Be Thankful road, a well-known scenic route, before returning home.

A forensic digital analyst went to work on Gilroy's phone. 'When you switch a mobile phone off,' explains Angus, 'it records the phone tower that it was last communicating with, so that when it's switched back on, it can quickly find it again.' On his way to the school in Argyll, Gilroy had switched off his phone between Stirling and Inveraray. Police suspected he had done this to avoid being tracked as he searched for a good place to dispose of Suzanne's body in the dense woodland. Then he went to visit the school. On his way back, Gilroy again switched off his mobile phone between Stirling and Inveraray. This, the police believed, was when he dumped the body.

When Gilroy stood trial, police search teams still hadn't found Suzanne's body. Nevertheless, on 15 March 2012, David Gilroy was found guilty of murder and conspiracy to defeat the ends of justice. The judge, Lord Bracadale, agreed to let television cameras into the court, making Gilroy the first convicted killer to have

Police searching for Suzanne Pilley's body near Arrochar, Scotland. Her remains were never found, though David Gilroy was found guilty of her murder in 2012

his sentencing filmed for British television. 'With quite chilling calmness and calculation,' said Bracadale, 'you set about disposing of the body, apparently somewhere in Argyll; and, but for the commendably thorough investigation carried out by Lothian and Borders Police, you might well have been successful in avoiding detection and prosecution.' He sentenced Gilroy to a minimum of eighteen years in prison. After receiving threats from fellow inmates at Edinburgh Prison, Gilroy was moved to Shotts Prison, where on his first day another inmate broke his jaw.

Gilroy's conviction had much to do with the sensitivity of investigators to his digital footprint. Without their analysis of mobile phone and CCTV evidence, he would probably be a free man today. It's rare for murderers to be convicted in the absence of their victim's body. It happened to Stuart Campbell, partly because of the splatter of Danielle's blood which investigators found on the underwear in his loft; and it happened to the Liverpudlian drug dealer caught out solely by DNA found in the pupal cases of maggots that had fed on his victim's corpse (see p. 57). In the Gilroy case there was no DNA. The scratches on his arm would not have been enough. He was convicted because of unusual mobile phone activity, CCTV video and images from road-side cameras.

It's up to people like Angus Marshall to use images and video to incriminate criminals like David Gilroy. The job is occasionally revelatory, usually methodical; it can take time to build up a digital picture. Angus creates his own tools to help. 'I'm a weirdo. I don't use any of the industry standard tools; they'd get me the same results as everyone else. Most of the programs I write are not very big or complicated, they simply automate things and allow me to sleep occasionally.' Once such programs have recovered all of the photographs and video files on a given hard drive, another goes through and tries to match them to a child abuse database held by the police, automatically sorting them into one of the five levels of severity, from relatively innocent nude posing right through to bestiality. 'Unfortunately there are always a few that haven't been seen before and some poor soul has to sit and manually classify those and then submit them,' Angus says, his genial expression clouding over.

The database stores the origin of each image, if it's known. This means investigators can link the consumers of illegal media to the creators, as happened in the busting of Scotland's largest paedophile ring in 2005 (see p.186). It's a traumatic job, but independent experts like Angus – or, more usually, police officers – look very carefully at abusive photographs and videos, to pick up clues as to where in the world they were taken. 'It can be subtle little things like the shape of the electrical sockets, the sound of the TV or the language being spoken,' Angus explains. 'You can approximate the time of day from where the sun is in the sky. If there is a victim of abuse in there, you can estimate their age and cross-reference what they look like against missing persons databases.'

And then there's the metadata – information which is embedded in images and video files taken on digital cameras and smartphones. Metadata reveals useful information, from the make and model of the device to the date and time when the media was recorded – if the perpetrator set the clock. Although image manipulation software and file sharing sites sometimes strip metadata, it is often still buried there and, with the right software, it can be read.

Modern devices even put GPS coordinates into the metadata, making it possible to know where the photographer was standing. This means digital forensic experts can interrogate the records of mobile phone networks to find out which phones were active in a particular area at any given time. GPS coordinates in metadata have also helped police locate criminals who are on the run, as demonstrated by the sensational case of John McAfee, a somewhat unstable computer genius who lived in the jungles of Belize.

McAfee was the son of an English woman who fell in love with an American soldier stationed in the UK during the Second World War. As a boy he moved with his parents to Virginia. When he was fifteen, his alcoholic and abusive father shot himself dead. McAfee then became hooked on drugs, but maintained an enthusiasm for computer programming and managed to hold down jobs at institutions as august as NASA. Eventually he struck out on his own and created McAfee Anti-virus, the first commercially available virus prevention software. In 1996 he sold his stake in the company for tens of millions of dollars. By then, as McAfee himself acknowledges, people knew him as 'the paranoid, schizophrenic wild child of Silicon Valley'.

In 2008, at the age of sixty-three, McAfee headed south from California to Belize, where he hoped to use the jungle flora to develop new antibiotics that would, in his words, 'interrupt bacteria's ability to communicate'. In 2012 police raided his research facility, claiming it was a methamphetamine factory. All charges were subsequently dropped.

Above: John McAfee surrounded by the media after his detainment by Guatemalan police

Left: McAfee's house in Belize

But the relationship between McAfee and his American expat neighbour, Gregory Faull, soured beyond repair. Faull, the owner of an Orlando sports bar, particularly hated McAfee's dogs. He sent a complaint to the local authorities, part of which read: 'These animals get loose and run as a pack. Three residents have been bitten and three tourists have been attacked.' McAfee later found four of his eleven dogs poisoned and had to shoot them to put them out of their misery.

On 11 November 2012 a housekeeper discovered Faull on his patio, lying face up with a bullet in his head. When police came to question McAfee, he hid from them under a box. Then he went on the run, disguised as a ragged salesman. However, he continued to update his blog and give online interviews. 'I have modified my appearance in a radical fashion,' he wrote; 'I'll probably look like a murderer, unfortunately.' When he made his way illegally over the border into Guatemala, the editor-in-chief of *Vice* magazine decided to follow his life on the run, and brought a photographer with him.

On 3 December the *Vice* website posted a photograph of McAfee in front of palm trees, underneath the smug caption: 'WE ARE WITH JOHN MCAFEE RIGHT NOW, SUCKERS.' But it also contained metadata which held clues to McAfee's exact longitude and latitude. Realising this, the photographer then posted on Facebook that he had manipulated the metadata. But that was a lie and soon the Guatemalan police tracked down and detained McAfee. He then faked a heart attack in order to buy his lawyer time. Together they blocked the Guatemalan authorities' attempt to deport McAfee back to Belize. He was sent to Miami, instead, where he was released. He then travelled to Montreal, Canada. Belize police still describe McAfee as 'a person of interest' in the murder of Gregory Faull, but not a prime suspect.

McAfee is now back in Silicon Valley, where he has been developing a $100 gadget called a D-Central that connects with your computer, smartphone or tablet and, McAfee promises, makes you invisible on the net. 'If you cannot see it, you cannot hack it, you cannot look at it, you cannot spy on anything happening inside it.' The idea is attractive to people in the light of the Edward Snowden leaks, and perhaps even more attractive to McAfee himself, given his own unhappy experience of overexposed data.

The D-Central is an extreme device for keeping private communications private, a manoeuvre that tech-savvy criminals and law-abiders alike are keener than ever to make. 'Certainly, younger generations are very careful about their footprints,' says Angus. 'I've spoken to a number of them over the years and they are fully aware of how much snooping goes on and how much of their personal data is being exploited. A lot of them have a simple solution to making sure people can't get at their data; they lie, they create fake accounts and leave fake footprints.' Some do this to stop potential employers seeing photos of them topless and drunk; others because they don't like the idea of government officials poring over their data; still others because they want to keep their criminal behaviour under the radar.

Angus is unhappy with the snooping activities of the National Security Agency (NSA) in the United States; in particular, he is unhappy with the notion of providing public security through imperilling individual privacy. 'We used to think Eastern Europe was bad. Our allies are getting even worse.' When an agency like the NSA snoops on websites like Google Mail or Facebook they use an automated program to look for trigger words. If you were to send an email to your lover saying, 'You're the bomb,' Angus reckons, 'they'd have a look at it, probably have a laugh, save it for the Christmas party but not much more than that. But if you start talking about building nuclear warheads, they'd look at you in great detail.' Of course many of the most serious criminals steer clear of providers like Gmail and Facebook.

Some of them also know that if they use the web on their smartphones or tablets via applications like the Facebook app, they will leave a trace that Angus can pick up. 'But if it's on a web browser on a mobile device then there is no trace. So we have to ask Facebook Inc., who do give us something. Twitter give us practically nothing.'

The big Californian corporations are trying to get everyone to put their personal data in 'the cloud', which ironically means a remote storage facility in the United States. The cloud keeps personal data up-to-date across all of a user's digital devices, thereby making it easier for the corporations to mine and exploit. Paradoxically, the more accessible the data is for users and corporations, the more hidden it is from people like Angus.

The future, says Angus, 'is online and it's cloud. Devices are pushing more and more of their data up into the cloud so that it is accessible to them everywhere. So we find it harder to get material off the devices, because it's not actually on them. We have to identify firstly if we can technically extract the data from the cloud and, secondly, if we can we get legal authority to do it.' Crossing international borders is just as difficult for a detective now as it was before cloud computing, but the need is far greater.

Angus recalls a recent case when a judge wrote to a social media company asking them two questions about the reliability of their user data logging. 'We got back a very simple response from that company's lawyers. It said firstly, "You have written to the wrong office. Don't write to us in America, write to us in Dublin." And secondly, "Under the terms of the treaty that exists between the United Kingdom and the United States, we don't have to answer your questions."'

Cloud computing presents other difficulties for the forensic expert. Software like Dropbox, which keeps files synced across devices, enables users to overwrite and change files on one device from another device anywhere else in the world. Angus calls this a 'massive benefit to the end user but, from an investigative point of view, if somebody has made a change on their computer in their house on this side of the country, and their laptop in the other house on the other side of the country is still switched on, Dropbox changes the content on the laptop, meaning I cannot tell which house you were in.'

If done deliberately, this kind of behaviour is known as 'anti-forensics', and it can take dozens of forms. A simple example is an organised criminal who buys a pay-as-you-go phone a couple of days before committing a crime, and then throws it away immediately after the crime. There are all sorts of more complicated anti-forensic techniques. Some programs allow users to change the metadata in files, so that they can make a file look as if it was created in 1912 and last accessed in 2050. Others make files look to forensic programs as if they were another kind of file altogether.

Thus, an expert could be tricked into thinking that an image file of a child being abused was an mp3 music file. Seeing through these ruses comes down to the ingenuity and experience of the forensic digital analysts. Just as a psychological profiler needs to empathise with a criminal in order to understand their motives and predict their actions, so a digital analyst has to be at the cutting edge of developments in the field, in order to work out exactly what tech-savvy criminals are up to.

Sometimes the experts dabble in anti-forensics themselves. Angus explain, 'I have colleagues who travel the world and they don't take any technology with them. They buy a new laptop and a new mobile phone in whatever country they are visiting and they trash it and leave it there.' They do this because airport staff in some countries routinely make sure that people aren't smuggling out the truth about what's happening in their country, or bringing in pornography or bomb-making instructions. They simply need access for a remarkably short time. 'All airport staff have to do is pull you off into the room where the member of staff with the rubber gloves can keep you busy for half an hour or so,' says Angus. And that's long enough for them to copy an entire hard drive.

In the case of cyber crimes like hacking, forensic digital analysts sometimes have to play catch-up with the criminals. The old adage holds true: When the forensic scientist takes a step, the criminal counters with a step of their own. Fingerprinting made burglars put gloves on. CCTV made kids pull their hoods up. So sometimes old technologies can be the best anti-forensic tools. Analogue cameras don't embed metadata into their photographs. Old-style bulletin boards can be set up in cyberspace and used completely under the radar. 'They are really easy to set up,' Angus reveals. 'The old software's still out there. The hardware is readily available now. It doesn't take a lot and, to be honest, you can hang one on the end of a pay-as-you-go mobile phone, which is almost untraceable.'

Physical evidence is still absolutely crucial to solving the vast majority of crimes. 'None of the cases that I've dealt with have been exclusively built on computer evidence,' Angus admits. 'The computer evidence is corroborating something else. It can be incredibly strong corroboration, but it's rare that it's the only evidence. So if we don't find it, as I said before, absence of evidence doesn't mean evidence of absence.'

ELEVEN

FORENSIC PSYCHOLOGY

'Every thief has a characteristic style or modus operandi *which he rarely departs from; and which he is incapable of completely getting rid of; at times this is such a distinctive feature that even the novice can spot it without difficulty, but ... only a practised, intelligent, and fervent observer is capable of distinguishing those traits, often delicate but identical, which characterise the theft, and drawing important conclusions from them.'*

Hans Gross, *Criminal Investigation: A Practical Textbook* (1934)

There's more to crime than breaking the law. Mostly, you have to want to do it. With very few exceptions, a criminal act cannot be punished without *mens rea*, that is to say, 'criminal intent'. In other words, if a lawbreaker didn't know what they were doing – because they are insane, say, or under the influence of mind-altering drugs – they will be given treatment rather than punishment for doing it.

Although it's often at the heart of crime fiction and dramas, motive is usually the least pressing concern in a real murder investigation. Hard forensic evidence, means and opportunity are the focus of such inquiries. But sometimes motive can be useful in pointing investigators in the right direction to look for that solid evidence.

Discovering that a missing child has made allegations of sexual abuse can turn a runaway inquiry into a far more serious operation, for example. And juries love motive because it helps them to make sense of events that are far outside their own experience of the world.

Motive-hunting is much more difficult when a criminal has victimised several people outside their immediate circle – so-called 'stranger' attacks. The motives of the serial killer may be amorphous, multi-stranded, developed over a lifetime, or in existence only for a nanosecond.

Psychologists generally agree that factors outside a killer's control, such as their upbringing and heredity, can have a defining effect on the way they behave as adults. Researchers have pursued several theories in their attempts to explain why some of us grow up to be serial murderers. Sometimes the answers they discover are profoundly shocking.

American neuroscientist James Fallon studied the brains of several convicted serial killers and noticed that many showed lower than average levels of activity in the areas of the frontal lobe linked to empathy, morality and self-control. In an attempt to quantify the differences between them and the general population, Fallon laid the brain scans on his desk, and mixed them in with scans he'd taken of his own family members. The scan that shouted 'Psychopath!' loudest was his own. He thought about burying the troubling result, but decided instead to play the good scientist and investigate further, by testing his own DNA. The outcome was even more unsettling. 'I had all these high-risk alleles for aggression, violence and low empathy.'

Seriously concerned by now, Fallon did some digging into his genealogy. In branches of his family tree he found seven alleged murderers, among them the source of the notorious rhyme:

Lizzie Borden took an ax,
Gave her mother forty whacks
When she saw what she had done
She gave her father forty-one.

Looking for answers to his own lack of criminality, Fallon decided he owed his non-violence to his mother's love and thanked her from the bottom of his heart. In 2013 he wrote a book called *The Psychopath Inside,* in which he says: 'It's not a death sentence, the biology, but it will give you some high potential for these things. The genes load the gun and make someone vulnerable to becoming a psychopath.'

Like Fallon, the first scientists to get involved in criminal trials wanted to identify abnormal minds. Medically trained, they were interested in the mental capacities of offenders and in trying to diagnose 'diseases of the mind'. When did an accused man have *mens rea*? When was he not responsible for his actions?

When the police came across bizarre crimes they couldn't make sense of, they began to seek help from those psychiatrists and psychologists with experience of mentally ill patients. The assumption usually was that the perpetrators of these perverse crimes were 'mad'. The guidelines for criminal insanity that are still a keystone test in many jurisdictions today were established in 1843 following the case of Daniel M'Naghten, who was acquitted of murder on the grounds of insanity after he shot dead the Prime Minister's private secretary, Edward Drummond. The rules can be summed up thus: Did the defendant know what he was doing, and, if so, did he know that it was wrong?

Sometimes, the criminal acts seem to leave little room for doubt. In 1929, Peter Kürten, also known as 'the Vampire of Düsseldorf', hammered, stabbed and strangled to death at least nine German children. While Kürten awaited his execution, Karl Berg, an eminent psychologist, won his confidence and got him to talk openly about his crimes. 'The sexual urge was strongly developed in me,' Kürten said, 'particularly in the last years, and it was stimulated even more by the crimes themselves. For that reason I was always driven to find a new victim. Sometimes even when I seized my victim's throat, I had an orgasm; sometimes not, but

Peter Kürten, the 'Vampire of Dusseldorf'

Below: Police searching the Pappendell Farm, Dusseldorf, for the bodies of Kürten's victims

then the orgasm came as I stabbed the victim. It was not my intention to get satisfaction by normal sexual intercourse, but by killing.' Kürten's weapon of choice was a pair of scissors. The sight of blood became ever more necessary for him to achieve orgasm. He even asked Berg hopefully if he'd be able to hear the blood gushing from his torso for a moment after the guillotine had sliced through his neck.

Possibly the most shocking thing for the people of Düsseldorf was that the 'vampire' who had terrorised their city did not look like a madman. 'He was slim and, comparatively, a good-looking man, with thick yellow hair always neatly parted, clever-looking blue eyes,' it was reported. When he arrived in court on the first day of his trial he was '[d]ressed in an immaculate suit ... with the look of a prim and proper businessman'. Nothing in Kürten's appearance or demeanour betrayed his nightmarish childhood of violence, marital rape and incest. Yet he appeared completely detached from reality in his extensive interviews with Berg, and at other points in his life. If he had not been like this, he would never have been able to befriend so many victims. So, although his crimes seemed to indicate madness, the man was less easy to draw a neat box around.

Despite the impossibility of narrowly defining one singular criminal mind in the way that Cesare Lombroso had tried to in the nineteenth century, by the time of Peter Kürten criminalists such as Hans Gross understood that there were a variety of criminal minds which could be partially read by the light of crime scene clues. The everyday behaviour of a serial offender tends to be in some way consistent with their criminal behaviour. For example, if a sexual murderer has had a partner before, they will usually have abused them (as Kürten had his wife). Forensic psychologists use this 'consistency principle' to build up profiles of serial offenders, which can help police focus their investigations.

Very probably the first 'offender profile' was written in 1888 in the midst of a slew of murders in Whitechapel, east London. At 3.40 in the morning on Friday, 31 August, a cartman was walking down Buck's Row when through the murk he made out a woman lying prone on the pavement, her skirt pulled up over her stomach. The cartman approached and found her hand cold to the touch. The only streetlamp was at the other end of the row, and the cartman couldn't be sure if she was drunk or dead. He pulled her skirt down to cover her modesty and went in search of a policeman.

Beat officer John Neil arrived at the scene to find blood seeping from the woman's throat. It had been slashed from ear to ear,

'Jack the Ripper' was a media sensation: here, a contemporary magazine cover depicts Constable Neil's discovery of Mary Anne Nichols's body

ferociously enough to sever the spinal cord. When the woman was brought into a 'deadhouse', Inspector John Spratling lifted up her clothes: her intestines were sticking out through a slit in her abdomen that went all the way to the breastbone. A reporter in *Reynolds' Newspaper* wrote: 'She was ripped open just as you see a dead calf at the butcher's.' The pathologist found two stabs to the woman's genitals and felt that the murderer 'must have had some rough anatomical knowledge, for he had attacked all the vital parts'. She was soon identified as Mary Ann Nichols, a 43-year-old prostitute. Most of her worldly possessions she had with her – a white handkerchief, a comb and a piece of mirror.

Over the next two and a half months, three more prostitutes were found murdered in the dark streets of Whitechapel. When a fifth, Mary Jane Kelly, was found slaughtered in a rented bed on 9 November, Scotland Yard were still no closer to identifying the killer, who had by now been nicknamed Jack the Ripper. In despair the police called in Doctor Thomas Bond, the police surgeon for the Westminster division, to assess the surgical skill of the murderer. The scene of Mary Kelly's death caused Bond's stomach to churn in horror. He couldn't find a heart inside her chest. The Ripper had taken it with him.

Later, in the calm of his office, Bond took a deep breath and tried to think carefully about what he had seen. First of all he answered the central question the police had posed. In fact, contrary to what the original pathologist had concluded, he decided that the murderer 'does not even possess the technical knowledge of a butcher or a horse slaughterer or any person accustomed to cut up dead animals'. But Bond wanted do more than state what the Ripper *wasn't*. He wanted to give the police positive guidance on who the Ripper *was*. He examined the police reports and post-mortem notes of the dozen or so prostitutes murdered in the previous seven months in Whitechapel, and decided that five were definitely carried out by the same man. The Ripper attacked between midnight and 6 a.m., with a long knife, in a one square mile area around Whitechapel.

The overkill actions – so-called 'signatures' – interested Bond as much as the basic details. The Ripper left his victims degradingly positioned on their backs with their legs apart, their guts out or missing and their throats cut. His degree of mutilation increased murder on murder: a classic example of confidence leading to an escalation in the level of violence. Four of his victims he had left out on the street. But his last one, Mary Kelly, had been killed indoors to give him more time and privacy to mutilate her. Bond described the Ripper as 'subject to periodical attacks of homicidal and erotic mania', and went on give his now-famous profile:

> A man of physical strength and great coolness and daring … The murderer in external appearance is quite likely to be a quiet inoffensive looking man probably middle-aged

> and neatly and respectably dressed. I think he must be in the habit of wearing a cloak or overcoat or he could hardly have escaped notice in the streets if the blood on his hands or clothes were visible ... he would probably be solitary and eccentric in his habits ... possibly living among respectable persons who have some knowledge of his character and habits and who may have grounds for suspicion that he is not quite right in his mind at times.

Some elements of the Bond profile were tenuous – why 'probably middle-aged'? – and it ignored other factors, such as the lack of semen at the crime scenes. Nevertheless, the report greatly influenced senior police officers and government figures involved in the investigation. Of course, because the police never caught Jack the Ripper, we can't know how accurate Bond's profile was. But it was a careful appraisal, peppered with important qualifying words still used in profiling today, such as 'likely', 'possibly' and 'probably', and it did address important issues such as how the Ripper escaped the scene of his crimes unnoticed.

By the Second World War offender profiles remained only a small part of police investigations, used for crimes that were difficult to comprehend. But some psychiatrists were happy to help, among them Dr James Brussel, Assistant Commissioner of Mental Hygiene for the state of New York. Brussel lived in the city's West Village, where he smoked a pipe and buried his head in Freud. He was no shrinking violet. One of his many books was called *Instant Shrink: How to Become an Expert Psychiatrist in Ten Easy Lessons*. His best-known forensic work involved profiling the 'Mad Bomber of New York', whose campaign lasted for sixteen years.

On 16 November 1940, a worker discovered a small pipe bomb filled with gunpowder on a windowsill at the energy company Consolidated Edison's offices in New York. Wrapped around it was a handwritten note: 'CON EDISON CROOKS – THIS IS FOR YOU.' The bomb was a dud. Ten months later a similar device was found in a street about five blocks away from Con Ed headquarters, also with a note. It, too, was a dud.

After the Japanese attacks on Pearl Harbor in December 1941, the New York police received a letter which read: 'I WILL MAKE NO MORE BOMB UNITS FOR THE DURATION OF THE WAR – MY PATRIOTIC FEELINGS HAVE MADE ME DECIDE THIS – LATER I WILL BRING THE CON EDISON TO JUSTICE – THEY WILL PAY FOR THEIR DASTARDLY DEEDS.'

Sure enough, New York was free of pipe bombs until 1951. But at that point the Mad Bomber began a renewed offensive. Over the next five years he planted at least thirty-one bombs, mainly in public buildings, including theatres, cinemas, libraries, railway stations and public toilets. Each bomb was a length of pipe filled with gunpowder and wrapped in a woollen sock, with a timer made from torch batteries and a pocket watch. Sometimes the police received warning calls; other times the bombs didn't explode; and sometimes notes reiterated that the campaign would continue until Con Ed had been brought to justice.

The first pipe bomb exploded in March 1951, near the Oyster Bar at Grand Central Station. At Loews Theater on Lexington Avenue in December 1952 one of the Mad Bomber's 'units' injured someone for the first time. In November 1954 a bomb stuffed into a seat at the Radio City Music Hall blasted through an audience watching *White Christmas,* injuring four. Six more people were hurt in December 1956 by a bomb at the Paramount Theater in Brooklyn, where 1,500 people were watching *War and Peace*. The city was in uproar. The New York Police Department (NYPD) launched the greatest manhunt in its history. They believed they were after a former employee of Con Ed with a grudge. But fingerprint examiners, handwriting experts and the bomb investigation unit had not been able to narrow it down further than that.

The NYPD called in Brussel. He studied the records of all the cases, examined the crime scenes and the bomber's methods, and developed what he called a 'portrait': 'By studying a man's deeds, I have deduced what kind of man he might be.' Brussel thought the Mad Bomber must be a skilled mechanic, of Slavic descent, a practising Catholic, living in Connecticut, over forty years old, neat, tidy and clean shaven, unmarried, and possibly a virgin. Warming to his task, Brussel noticed that in his handwritten letters the bomber had produced 'w's that resembled two 'u's, rounded like a pair of breasts – he therefore must not have developed beyond the Oedipal stage of psychological development, and was probably living with a mother figure, such as an older female relative. Brussel thought the bomber was suffering from paranoia, and concluded with a precise prediction: he would be wearing a buttoned-up double-breasted suit when the police arrested him.

At Brussel's request, the profile was published in *The New York Times* on Christmas Day 1956. This was probably his greatest contribution to the capture of the bomber. On Boxing Day the *New York Journal-American* published an open letter promising the bomber a fair trial if he gave himself up. He replied that he wouldn't, and listing his grievances against Con Ed: 'I was injured on a job. My medical bills and care have cost thousands ... I did not get a single penny for a lifetime of misery and suffering.'

This response led Alice Kelly, a company clerk, to look at Con Ed's pre-1940 employment records – which the company had previously told police were destroyed. There, Kelly found a file on George Metesky, who had worked for Con Ed between 1929 and 1931 as a generator wiper, and who had been injured in an accident at the 'Hell Gate' plant. Metesky had inhaled a gust of gas which he claimed had damaged his lungs, leading to pneumonia and tuberculosis. He was fired from his job without compensation, which led him to write 900 letters to the Mayor, the Police Commissioner and the newspapers. 'I never even got a penny postcard back,' he later said. Looking through Metesky's letters of complaint, Kelly noticed that several of them included 'dastardly

Police lead away George Metesky, the 'Mad Bomber of New York'. Here he is wearing the double-breasted suit Dr James Brussel, who provided a psychological profile of the bomber before his arrest, specified that he would be wearing when found

deeds', the same dated phrase that the Mad Bomber had peppered his notes with.

On 21 January 1957, the police arrived at Metesky's address in Westchester, Connecticut. He opened the door wearing his pyjamas, having just settled down to spend the evening with his two older sisters. His sisters told the police that he was an impeccably neat man who went to Mass regularly. When he came back downstairs from getting dressed, Metesky was wearing a buttoned-up double-breasted suit. He told the police that he never wanted to hurt anyone, and had designed the bombs accordingly. A doctor declared Metesky insane and unfit to stand trial, and he was committed to the Matteawan State Hospital for the Criminally Insane. He was released in 1973 and died twenty years later, at the age of ninety.

Despite the legend of Brussel's profile, it was Alice Kelly's meticulous searching of the records armed with the clues from his letters of complaint that nailed Metesky. But the Brussel profile was hailed as a piece of interpretative genius because it correctly depicted the bomber as a paranoid Catholic Slav living in

Connecticut and wearing a particular type of suit. His deductions were logical, not magical: bombing is a crime associated with paranoia; in the post-war years protest bombings were common in eastern Europe; most Slavs were Catholic; lots of Slavs lived in Connecticut; and the fashion in the 1950s was for men to wear double-breasted suit jackets with the buttons done up.

The most shocking aspect of the case is that it took the NYPD sixteen years to track down the Mad Bomber, even though he had given them so many clues in his notes: 'I AM NOT WELL, AND FOR THIS I WILL MAKE THE CON EDISON SORRY.' Malcolm Gladwell concluded in a 2007 *New Yorker* article that 'Brussel did not really understand the mind of the Mad Bomber. He seems to have understood only that, if you make a great number of predictions, the ones that were wrong will soon be forgotten. The Hedunit is not a triumph of forensic analysis. It's a party trick.' But there were no such detractors at the time. There was only relief. Brussel's profile played a large role in subsequently encouraging the police to call on psychologists and psychiatrists to deliver profiles for their investigations into serious crime.

In 1977 the FBI inaugurated profiling training courses at their academy in Quantico, Virginia. They were the brainchild of Howard Teton, who acknowledged James Brussel as 'a true pioneer of the field' and was very influenced by what he saw as Brussel's successes. A small group of FBI agents drove off on weekend trips to jails, where they interviewed thirty-six serial killers and serial rapists. They wanted to base their future profiles on empirical evidence rather than hunch and anecdote. Their research produced two models of serial killer: the disorganised man, who attacks victims at random, not caring who they are, murdering them sloppily, and leaving forensic traces; and the organised man, whose victims fulfil a specific personal fantasy. He takes his time with them, and rarely leaves forensic traces.

Placing serial killers within such binary categories is alluring – and enduring – but it's more accurate to put them on a spectrum. Whilst some are always disorganised, others become more organised with time. Jack the Ripper, for example, dealt with Mary

Kelly, his fifth and probably final victim, in the privacy of a rented room, the better to mutilate her. And escalation doesn't always make killers more organised. As their need for violence and blood increases, their attacks may become more disordered and careless. Because of Hollywood we are quite accustomed to thinking of serial killers as enigmatic, very clever, white and middle-class. This is partly supported by the data: according to statistics, they tend to be slightly above average intelligence, single, white and (with some notable exceptions) working- or middle-class.

And as forensic scientist Brent Turvey has pointed out, 'You've got a rapist who attacks a woman in the park and pulls her shirt up over her face. Why? What does that mean? There are ten different things it could mean. It could mean he doesn't want to see her. It could mean he doesn't want her to see him. It could mean he wants to see her breasts, he wants to imagine someone else, he wants to incapacitate her arms – all of those are possibilities. You can't just look at one behaviour in isolation.'

For many of us, our first encounter with the very concept of a criminal profiler would have happened in the dark. The 1991 film *The Silence of the Lambs*, based on Thomas Harris's compelling novel, introduced us to FBI agent Clarice Starling, played by Jodie Foster. Trainee agent Clarice is chosen for a serial killer task force because her bosses believe she'll be able to elicit help from Hannibal Lecter, a brilliant forensic psychiatrist who is incarcerated for a series of cannibalistic murders. Both the film and the book weave an intricate web of riddles and red herrings, encapsulating the difficulties of profiling a serial killer.

Thomas Harris's Hannibal Lecter novels were among the first to latch on to the idea of offender profiling, and it has since proved to be fertile ground for crime writers, myself included. For the writer of fiction, understanding the motivation of our characters is at the heart of what we do; the forensic psychologist offers us the perfect fantasy figure – someone who looks at people with an analytical and empathetic eye, but who also gets to be the hero.

But it wasn't just us writers who were intrigued by the possibilities of offender profiling. In the mid 1980s, police forces around the world were already fascinated by the 'offender profilers' the FBI were training up. They offered fresh hope in cases that seemed to have dead-ended.

For four years, the Metropolitan Police had been trying to track down a rapist who had been violently attacking women in London. The attacks began in 1982, when a man in a balaclava raped a woman near Hampstead Heath Tube station. More rapes in north London followed in similar circumstances. On 29 December 1985 the 'Railway Rapist' became the 'Railway Killer' when he dragged 19-year-old Alison Day off a train, gagged her, bound her, raped her and strangled her with a piece of string.

By this point the police had linked the same man – who sometimes struck with an accomplice – to forty rapes. Then a 15-year-old Dutch girl, Maartje Tamboezer, was attacked as she cycled through woodland near a railway station in Surrey. Two men dragged her for half a mile before raping her, strangling her with her own belt and setting her body on fire. Only a month later, local TV presenter Anne Locke was abducted and murdered as she got off a train at Brookmans Park in Hertfordshire. The suspect list had reached unmanageable proportions. A fresh approach was needed.

In 1986 the Met contacted David Canter, an environmental psychologist at the University of Surrey. They had a single question: 'Can you help us catch this man before he kills again?'

All the attacks had taken place at night on or near railway stations, the victims were usually teenaged girls, who were raped and then, in three cases, garrotted to death. Canter looked at the dates and details of the attacks, and plotted their locations on a map. He suggested that the rapes had begun opportunistically but had become increasingly planned. He thought the culprit had committed the early crimes in the area that he was familiar with, close to home, and then ventured further afield where he wouldn't be recognised. Based on witness statements and police reports Canter built up a profile of the personality and lifestyle of the masked attacker. He suggested that he was married but without children (because he talked to some of his victims normally before he attacked them); that he had a semi-skilled job (based on his ability to plan the later crimes); was in his twenties (from witness reports); and 'probably had a history of being violent against women and was quite a nasty character and would be known to be so'.

On the basis of Canter's profile the police started following John Duffy, a carpenter who had spent some time working for British Rail, and who lived very close to the first three attacks, in Kilburn. Duffy was on the police's suspect list because he had raped his estranged wife at knifepoint. But he was low down because some officers had thought that 'just a domestic'. When Canter contended that the Railway Rapist would have a history of this kind of violence, Duffy was pushed up the list. The police arrested him while he was following a woman in a park, and strong forensic evidence linked him to two of the murders and four of the rapes. He was convicted in February 1988.

Thirteen of the seventeen points in Canter's profile turned out to match Duffy. He'd said that Duffy would be small (he was five feet four inches); feel unattractive (pockmarked by acne); interested in martial arts (spent a lot of time at a martial arts club and collected kung fu weapons); have souvenirs of his crimes (thirty-three of his victims' door keys). After Duffy's conviction, it became commonplace for UK police forces to ask psychologists to provide offender profiles for major crime investigations.

The only downside to the successful prosecution of Duffy was that his accomplice was still at large. For nearly ten years Duffy refused to talk about him. But forensic psychologist Jenny Cutler eventually drew the information out of him. A source said, 'He grew to like her. He was a social inadequate in a hostile male environment. He was struck by her in a way.' He finally revealed the name of his accomplice – childhood friend David Mulcahy. Both from working-class Irish backgrounds, the boys had been bullied at school, and had turned to each other. At thirteen

Mulcahy had been suspended from school for bludgeoning a hedgehog to death in the playground. Teachers found Mulcahy covered in blood with Duffy next to him, laughing. They committed their first rape together when they were twenty-two. In court at Mulcahy's trial, Duffy explained, 'We normally travelled by car. We called it "hunting". Part of it was looking for a victim, finding her and tracking her. David had a tape of Michael Jackson's "Thriller". We used to put that on and sing along to it as part of the build-up ... We did it as a bit of a joke, a bit of a game. It added to the excitement ... You get into the pattern of offending – it is very difficult to stop.' Combined with LCN DNA evidence which hadn't been possible at the time of the crimes, the evidence was incontrovertible. In 1999 Mulcahy was convicted of three murders and seven rapes, and Duffy of a further seventeen rapes.

The single most useful thing about David Canter's profile was its prediction of where the attackers lived. Before Duffy's conviction, Canter had been an environmental psychologist. After it he rebranded himself an 'investigative psychologist', and devoted much of his time to researching and writing about geographical profiling. Just as law-abiding citizens tend to go back to the same street to do their shopping time and again, most criminals like to commit their crimes in the same areas. They feel safer in places they know. David Canter came up with a circle hypothesis: if you draw a circle with a circumference going through the sites of the two crimes furthest apart from each other, the culprit's home will likely be near the centre of that circle. Research has shown this to be true of the majority of criminals who strike more than five times. Canter has found that a serial killer can usually be found living within a triangle formed by the sites of his first three murders, as Duffy was. He has developed a computer program, called the Dragnet, which generates 'hotspots'. Rather than attempting to mark a killer's home with an 'X', Dragnet produces areas showing where he is likely to live, colour-coded from the highest probability to the lowest.

My own close encounter with the use of computer algorithms in tracking serial offenders came courtesy of Kim Rossmo, a detective with the Vancouver Police Department. He was the first police officer in Canada to earn a doctorate in criminology and the research he conducted for his dissertation led him to develop a program that could predict where serial offenders lived. When we met, his system was being beta-tested by burglary investigators who were staggered by the results. I was so impressed by what I saw and heard that I used it as the basis for a thriller, *Killing the Shadows*, published in 2000, when the idea of geographic profiling was still in its infancy. Years later, I was on a book promotion tour in America when I turned on the TV one morning to see Kim Rossmo being interviewed during the hunt for the Washington Sniper. The bleeding edge had become mainstream in a few short years.

By the time I wrote *Killing the Shadows*, I had already published two novels featuring clinical psychologist and offender profiler Dr Tony Hill. When I first had the idea for his debut, *The Mermaids Singing*, I knew I needed help. In the UK, we did things differently from the FBI and the Royal Canadian Mounted Police. We didn't train cops in behavioural science; we brought practising clinicians and academics in to work alongside experienced detectives. I realised I had no idea how this worked in practice, or what a criminal profiler actually did. The man I turned to for help was Dr Mike Berry. And although I pillaged his working methods, let me say for the record that his personality is different from Dr Tony Hill's in every significant respect!

Like David Canter, Mike Berry is a psychologist who became involved in offender profiling just as the UK police started taking it seriously. He worked for many years at the sharp end, treating patients in secure mental hospitals before he turned to teaching forensic psychology at Manchester Metropolitan University. Nowadays he's based in Dublin, at the Royal College of Surgeons.

'I did my clinical training and undertook placements in clinical departments working with adults, people with learning disabilities and children, and in neuropsychology, before undertaking an elective six months in Broadmoor, working with Tony Black and colleagues.' Broadmoor is a high-security psychiatric hospital in Berkshire which, since its opening in 1863, has housed the most dangerous criminals in Britain, including Charles Bronson, Ronnie Kray and Peter Sutcliffe a.k.a the Yorkshire Ripper. Years later Mike moved to Ashworth Hospital in Merseyside, where he worked with some patients displaying extreme behaviour.

Having started their careers in the same era, Mike Berry acknowledges that David Canter's early forensic work was instrumental both in bringing two murderers to justice, and in boosting the art of geographical profiling. But he sees a flipside: 'It was too good. It went off too quickly. The press got on to it and then the police fell under a lot of pressure. The media would say, "You've had seven days and you haven't found somebody? When are you going to bring in the experts?" It built up an expectation that you knock on a door of the psychologist and in two hours they'll solve your murder.'

But then came the case that seriously undermined the public's trust in offender profiling. On 28 July 1992, the Met Police approached profiler Paul Britton. They needed help catching the perpetrator of an horrific crime committed two weeks earlier on Wimbledon Common, in south-west London. Rachel Nickell, a 23-year-old model with blue eyes and blonde hair, had taken her dog for a morning stroll, along with her two-year-old son Alex. She was passing through a lightly wooded area when a man jumped out at her and viciously stabbed her forty-nine times. In his autobiography *The Jigsaw Man* (1998) Paul Britton describes how Rachel was found 'in the most degrading position the killer could manage in the circumstances, with her buttocks prominently displayed ... her throat so severely cut that it appeared her head had almost been severed'. Alex was muddied but unhurt. When the next person walked through the trees and came across him, he was crying, 'Mummy, wake up.'

CSIs found a single shoe print close by Rachel's body, but no semen, saliva or hair belonging to her killer. Eyewitnesses reported seeing an average-looking man about twenty or thirty years old washing his hands in a nearby stream just after the murder. The media stoked tremendous interest in the case, and one local women's group offered to donate £400,000 to help the police with their investigation – although they couldn't accept it.

The police asked Britton to draw up an offender profile. He believed that the killer was a stranger because he wouldn't have wanted to risk Alex recognising him. He thought he would 'have a history of failed or unsatisfactory relationships, if any … Be likely to suffer from some form of sexual dysfunction, like difficulty with erection or ejaculatory control …' Because of the frenzied, disorganised nature of the attack and the lack of an attempt to hide the body, 'he would be of not more than average intelligence and education. If he is employed he will work in an unskilled or labouring occupation. He will be single and have a relatively isolated lifestyle, living at home with a parent or alone in a flat or bedsit. He will have solitary hobbies and interests. These will be of an unusual nature and may include a low level interest in martial arts or photography.' At the bottom of his report Britton left a warning: 'In my view it is almost inevitable that this person will kill another young woman at some point in the future as a result of the strong deviancy and aggressive fantasy urges already described.' It was in many respects a generic profile, which could have fitted a relatively large number of men.

Within a month of the murder the police had received more than 2,500 phone calls from the public, and were drowning in paperwork generated by the case. They used Britton's profile to narrow down the suspect list. When BBC *Crimewatch* featured a reconstruction of the murder, which included an edited version of the profile, three different callers put forward the name of Colin Stagg, 23, who lived by himself on an estate less than a mile from Wimbledon

Common. He had told a neighbour that he'd walked through the lightly wooded area ten minutes before Rachel was killed.

In September, when the police went round to Stagg's flat to bring him in for questioning, they were confronted by a sign on his front door – 'Christians keep away, a pagan dwells here'. Inside they discovered pornographic magazines and books on the occult. They interviewed Stagg for three days. When they asked him which shoes he had been wearing on the day of the murder, he said that he'd chucked that pair away two days before his arrest. He had had relationships with a few women but he 'just couldn't get it up' with any of them. In the days following Rachel's murder, he told police that he'd lain out on Wimbledon Common completely naked except for a pair of sunglasses, opening his legs and smiling at a woman who passed by. Stagg repeatedly denied murdering Nickell or being the man seen washing his hands in the nearby stream.

Stagg fitted Britton's profile closely and so became the police's prime suspect. But they didn't have enough evidence against him. They went back to Britton to see if there was anything he could suggest that could help them build their case. The stratagem that evolved was to use an attractive undercover policewoman in a 'honeytrap' sting.

Britton trained the police officer, known as 'Lizzie James', in several one-on-one sessions. She was to let Stagg know she was open to things that other people weren't and give him space to talk about whatever he wanted. Eventually she should tell him that as a teenager she had been lured into an occult group where she was abused and forced to watch the sexual murder of a young woman and child. Since leaving the group, all her relationships with men had failed because none of them had been potent or commanding enough to realise her fantasies for her.

Lizzie wrote to Stagg, who responded immediately. She sent him a photograph of herself and their correspondence gathered pace, with Lizzie encouraging Stagg to relate his fantasies to her:

> You asked me to explain about how I feel when you write your special letters to me. Well, firstly, they excite me greatly but I can't help but think you are showing great restraint, you are showing control when you feel like bursting. I want you to burst, I want to feel you all powerful and overwhelming so that I am completely in your power, defenceless and humiliated.

Stagg replied:

> You need a damn good fucking by a real man and I'm the one to do it ... I am the only man in this world who is going to give it to you. I am going to make sure you are screaming in agony when I abuse you. I am going to destroy your self-esteem, you will never look anybody in the eyes again ...

On 29 April they spoke on the phone for only the second time and Stagg told her a story, in which he entered Lizzie from behind and yanked her head back with a belt. The next day he sent a letter in which he admitted that he had been arrested on suspicion of the Nickell murder. 'I am not a murderer,' he added, 'as my belief is that all life from the smallest insect to plant, animal and man is sacred and unique.'

Five months after the correspondence had begun, Stagg and Lizzie met for the first time, in Hyde Park. She gave him the full account of her occult experience, and Stagg gave her a brown envelope. In it was a vivid fantasy involving Stagg, another man, Lizzie, a stream, a woodland, pain and a knife dripping with blood. At the end Stagg explained that he had written the story because he thought Lizzie would be 'into it'. The police were excited by this development and Britton told them, 'You're looking at someone with a highly deviant sexuality that's present in a very small number of men in the general population. The chances of there being two such men on Wimbledon Common when Rachel was murdered are incredibly small.'

In August 1993 the police arrested Colin Stagg. Over a year later, when the case finally came to court, Mr Justice Ognall reviewed the 700 pages relating to the case and took a dim view of the trap that the police and Britton had laid for Stagg: 'This behaviour betrays not merely an excessive zeal but a substantial attempt to incriminate a suspect by positive and deceptive conduct of the grossest kind. The prosecution sought to persuade me that the object of the exercise was to afford the accused an opportunity either to eliminate himself from the inquiry or implicate himself in the murder. I am bound to say I regard that description as highly disingenuous.' Ognall ruled the letters and taped conversations inadmissible, and Stagg walked free.

In 1998, Lizzie James took early retirement at the age of thirty-three, because of the post-traumatic stress she had experienced since the investigation. In 2002, Paul Britton faced a public disciplinary hearing before the British Psychological Society for offering advice to the Rachel Nickell investigation not backed by accepted scientific practice, and for making exaggerated claims about the effectiveness of his methods. But, after two days, the committee dismissed the case, judging that the eight years that had passed were too long for Britton to get a fair hearing. In those two days the committee also heard that the honeytrap had been approved at the highest levels of the Met Police, and that Britton's work had been checked by the FBI profiling unit in Quantico, Virginia.

In the same year the police established a cold case review team to look into Rachel Nickell's murder. Scientists re-examined Rachel's clothes and were able to produce a DNA profile with the help of a new extra sensitive technique (see p.153). The DNA did not belong to Colin Stagg. It belonged to Robert Napper – a paranoid schizophrenic who had raped as many as eighty-six women across London, before being caught and locked up in Broadmoor. In November 1993, sixteen months after Rachel Nickell's murder, Napper had brutally killed Samantha Bisset and her daughter Jazmine, 4, in their flat in Plumstead. On 18 December 2008 Napper was convicted of killing Rachel, too.

Forensic pathologist Dick Shepherd did the autopsies on both Rachel and the Bissets. He says that, when he was performing the

Bisset autopsies, he remembers remarking '"This guy has been here before, whoever did this, this is not the first murder and you are looking for a nasty – what about the Nickell killer? This looks like a progression," and everybody went, "Oh no, we've got Stagg for that, we keep an eye on him twenty-four hours a day."' When asked if the murders might be linked, Paul Britton called them 'a completely different scenario'.

The police had searched Napper's home in May 1994 and discovered his rare pair of Adidas Phantom trainers. It was not until a decade later that they matched these to the shoeprint left beside Rachel Nickell's body on Wimbledon Common. In December 2008, an editorial in *The Times* concluded, 'The reluctance to investigate Napper for the Nickell murder can be explained only by a belief, shared by the police, Paul Britton and the Crown Prosecution Service lawyers, that they already had their man. In their view Colin Stagg was guilty, so they ignored the material on Napper.' Stagg was a lonely man desperate to lose his virginity to a beautiful woman. The most explicit sexual story he wrote probably bore such a strong resemblance to Rachel's murder because he felt Lizzie James was into violent sex and used a local murder whose site he knew as inspiration.

Apart from being a major tragedy for the Nickell and Bisset families, the botched investigation was an expensive embarrassment for the Met. On top of the overall cost of the operation, Colin Stagg was awarded £706,000 in compensation (partly because his name had been so badly dragged through the mud that he never got a job again). Today, profilers are called Behavioural Investigative Advisers (BIAs) and must be accredited. Tellingly, the first guideline for a BIA with the Kent Police is that they '[k]now the limits of their expertise and discipline and stay within them'.

In contrast to David Canter's profile of the Railway Killer, the most unhelpful assertion in Britton's profile was that the Nickell murderer 'will live within easy walking distance of Wimbledon Common and will be thoroughly familiar with it'. In fact, Robert Napper had only very recently been forced there after police closed in on his usual hunting ground around Plumstead.

Mike Berry makes a point of visiting the crime scene at the same time of day as the offence was committed because he believes it helps him make tentative judgements about the offender's relationship to the site of his crime. He says, 'I remember years ago going down to visit the scene of a crime in a town park. The taxi driver gave me a torch and said, "I'm not letting you go there on your own, you'll never come out." It was midnight and pitch black in the park and this was influential in my profile, and I said, "Oh right, that's all I need to know." The body was found in a pond in the middle of the park. It became obvious that the person who killed the woman had to be very local to be able to take her there. Daylight pictures wouldn't have told me how pitch dark it was.'

Early gathering of simple information can provide the sturdy building blocks for a safe offender profile. When there is little forensic evidence to go on, local knowledge becomes even more important. 'I remember one case,' says Mike. 'We talked to the local beat officer and he said that in order to get back from the nearby city nightclubs, youngsters would get a taxi to the top of the woods. Then they walk down a path through the woods, stop in a clearing and have a drink and a cigarette there, and then walk on to the village where they live. He said to take a taxi all the way around the woods to the village will cost them twice as much. So the victim – a 16-year-old girl – wouldn't have been at all worried about walking with somebody in the woods, because that's what they did. The beat officer didn't know the importance of what he was saying but he had explained the atypical behaviour. Because the victim was in jeans and a shirt and there was no evidence that sex had taken place, it was suggested that she had rejected his advances and he lost his temper, grabbed her by the throat, strangled her and then walked off home.' This indicated – along with other factors – that the impulsive, unplanned killing was likely to have been committed by a young male living, or staying over, in the village. It is worth noting that when the police bring in the murder squad, they are all strangers to the district, so talking to the local beat officer will tell you much of the information. The police found the suspect within hours in the village. He fitted the profile completely and was later convicted.

From the first time we met, I've been fascinated hearing Mike talk about how he profiles. The method of profiling I've given to Tony Hill has its roots in the way he does things. The shelves of Mike's office are bursting with books about forensic psychology, including memoirs of profilers detailing their experiences. Mike is very aware of the pitfalls that some of them have fallen into. 'You've always got to say this is the characteristics of the likely killer, not the characteristics of the killer. Psychologists should not attempt to identify a specific individual as the killer.'

Mike's profiles are based on both empirical studies of offenders, and on his years of experience working with offenders in therapeutic and investigative settings. This gives him rich background material for his profiles which, like Dr Thomas Bond in 1888, he qualifies with 'likelys', 'possiblys' and 'probablys', unless he is absolutely certain of something.

Once Mike has visited the scene of the crime, he studies photographs and police notes, witness accounts, the autopsy report and photographs, and any other relevant information he can get his hands on. At this stage, it's important that the police keep suspect details to themselves, so the profiler is not influenced by the investigators' preconceptions. The most valuable profiles have to avoid any kind of bias or prejudice.

For Mike, the next stage happens inside his head. 'I sit at a blank screen and think, as part of developing the initial hypothesis to build up a profile. I go for walks and talk it over in my head and occasionally, if there's somebody I work with who I trust, I bounce ideas off them. Then it goes to the rejection model. OK, can we assume that it is a man? And of course nowadays it's becoming more common to see female killers ... I see what I can put in there and what I can reject ... If it's a sexual offence I'm looking at suspects from possibly ten years up to sixty years old. But they are likely to be slightly older than young teenagers who are having sex for the first time. You start with very crude things. If a condom is used then you think, "Why?" Because he has criminal knowledge, criminal experience, and he doesn't want to leave evidence ... I set up a model and the whole time I am looking at trying to reject it, asking myself, "Where's the evidence for that?" You can work on something for several hours and then something says, "No," so you throw it out. I think sometimes police officers and profilers make mistakes when they have a hunch and they stick with it. What we all have to do is to learn to let go. If the evidence doesn't support a hypothesis, you throw it away, start on Plan B and work your way through to Plan Z.' The profile would concentrate on a number of characteristics such as gender, age, racial group, occupation, relationship status, vehicle type, hobbies, criminality, relationship with females, with the victim, choice of victim, social class, education, post-offence behaviour, interview behaviour and so on.

Once they have looked at all the information about the crime, some profilers have a pre-prepared set of question to ask themselves. David Canter asks: What do the details of the crime indicate about the intelligence, knowledge and skills of the offender? Does he seem spontaneous or methodical? How has he interacted with the victim and does that tell me how he may have interacted with others? Does it seem like the criminal was familiar with the crime he committed or the scene he committed it in?

Rather than trying to find one man, like Colin Stagg, the goal for profilers should be to give detectives a report that makes the number of potential suspects more manageable. As an example, the Yorkshire Ripper inquiry gave investigators 268,000 named suspects to look into, and led them to make 27,000 house visits. Mike Berry says, 'If we've got a sexual murderer then we've got about 30 million men. If we cut the higher age and lower age we can bring it down to about 20 million ...' Anything that a profiler can do to safely bring that figure down further is very valuable for the police. Mike says, 'Some people still have a false image of profiling. They think that the profiler is going to come along and say he's left-handed, ginger-haired, five foot six, and supports [Manchester] City. But more people see it now as a tool in the detective tool bag, similar to DNA and pathology. It's very much a tool rather than a major source of information, and I think that's right.'

Does the glamour and thrill set in motion by James Brussel and Clarice Starling live on? 'It's a challenge but it's also very draining because you are dealing with sometimes very horrific offences. Certainly when I first started profiling I kept thinking it's my fault if I didn't get the information to the police that led to the right guy, but after a while you realise all you can do is say "here are the likely characteristics", it's still up to the police to collect the data and capture the criminal. Nowadays many detectives contact Bramhill, the police college, directly. The police have become more self-contained and more likely to use their own people. You rarely get psychologists doing it publically now.'

Of course, forensic psychologists do much more than help in the hunt for killers. Most work with institution-based offenders and patients, with some undertaking work in the criminal and civil courts. Mike Berry reckons he and his colleagues seldom do more than five court appearances each per year (although they may do a hundred or so reports in the same time frame, of which 5 per cent are contested, resulting in a court appearance to defend it). Outside the courtroom, forensic psychologists do a lot of work with criminals in Secure Units and psychiatric hospitals, sometimes trying to help them prepare for life outside, and other times trying to get them to provide information that can close other cases, as Jenny Cutler did with John Duffy. Mike says, 'I work with offenders and victims. It's very challenging. You have the challenge of trying to break down a story, trying to make sense of it ... sometimes you don't get the full story for several months or even years.'

Mike Berry is quick to praise the input of academic psychologists as well: 'Academics tend to be much more likely to ask "where is the evidence?" They are undertaking research in areas as diverse as speech analysis in rapists, movement in serial offenders, interviewing – all of which is very useful and we can extract some of the good points and use that in profiling. Interviewing offenders, victims and witnesses to get their full story is the area that David Canter, who Berry describes as 'the major force in the field', believes psychologists have contributed to most in crime investigation. An interview is all about asking someone to recall things, which is a notoriously tricky and error-prone activity. Psychologists have studied the art of the interview and have come up with some key points for detectives to follow: They need to build an open atmosphere with their subjects, recreate the context around the events in question, ask open questions which can't be answered with 'yes' or 'no', avoid cutting off the subject in mid flow, and be interested in every point they make even if it appears irrelevant. Although everyone is different, encouraging interviewees to tell the truth often comes down to the elusive qualities of trust and respect. The other proven technique to improve your chances of getting a suspect to confess is to make sure that they are fully aware of the weight of the evidence against them. But, contrary to the movies, a coercive approach can lead interviewees either to shut down or to make false confessions. And, of course, in these days of recorded interviews, any attempt at coercion will see the evidence thrown out in the courtroom.

Forensic psychologists are also increasingly involved in 'psychological autopsy', an attempt to discover the state of mind of someone before they died. A pathologist may have established the cause of death with a physical autopsy, but will not necessarily know whether it was suicide, murder or an accident. Psychologists look at diaries and emails, online activity, the mental health history of the deceased's family and may conduct interviews with people who were close to them.

In 2008, Mike Berry commented (for Sky Television) on the extraordinary case of the disappearance of schoolgirl Shannon Mathews, in Dewsbury, West Yorkshire. His analysis of events relied on exactly the sort of sensitivity to nuances of expression and behaviour which are needed for a psychological autopsy. 'I noticed that when her mother [Karen] was being interviewed on the sofa with her young partner, one of her children was trying to climb up on her lap and she kept pushing the child away. I thought if you've just lost one of your children the expected reaction is to hug the others tightly to you, and she didn't. Then she said something about "The street will be pleased when they find her", rather than "*I* will be pleased. I'll be over the moon."' It turned out that Karen had drugged her 9-year-old daughter with temazepam and given her to an accomplice, who had kept her for a month in his nearby house. The plan was for Karen's boyfriend to 'find' Shannon and then split the reward money with Karen. But, following a tip-off, police found the little girl in the accomplice's house, bundled into a drawer under the accomplice's platform bed.

A more conventional psychological autopsy was carried out following the death in 1976 of Howard Hughes, the eccentric American businessman who, at the age of eighteen, inherited his father's family business in Houston, Texas. By the age of sixty, Hughes was the world's richest man, but had developed a terror of infectious diseases. He moved to Mexico, where he injected himself with codeine, wore no clothes, let his hair and nails grow, never took a bath or brushed his teeth, and would spend up to twenty hours at a time sitting on the toilet. Because of his reclusive and bizarre behaviour, his will was challenged, and this led to Dr Raymond Fowler, president of the American Psychological Association, preparing a report to determine whether he was psychotic, and had therefore lost touch with reality. Fowler concluded that, although Hughes was mentally disturbed and extremely eccentric, he always knew what he was doing and was not psychotic. His will was accepted.

In his 1827 essay 'On Murder Considered as One of the Fine Arts', the writer Thomas de Quincey playfully suggests that murder should be examined from an aesthetic perspective, rather than a legalistic one. In a sense, that's what the forensic psychologist is doing. She (85 per cent of forensic psychologists are female) is trying to paint a picture that makes sense of the inside of someone's head. It may be a long way from beautiful, but it has meaning for the person who is home to those thoughts. And the better we understand these strange universes that our fellow human beings occupy, the closer we may come to fixing them before they leave a trail of destruction in their wake.

TWELVE

THE COURTROOM

'Unlike the prosecution, which has to prove its case, the defence merely has to introduce doubt in order to win'

Tim Pritchard, *Observer*, 3 February 2001

Working as a lawyer for thirteen years, Fiona Raitt treated scientific evidence as just another 'part of the process'. But when she went back to university, in Dundee, she started talking to scientists and psychologists about 'how it's collected from the crime scene, how it's stored, how it's used and eventually how it ends up in court'. Now, as Professor of Evidence and Social Justice, she writes about the tensions at play in each of those evidential steps: 'Everyone has a different vested interest in how the science is used, from the earliest stage of its discovery right up to its appearance in court.' The police may look particularly hard at evidence when they think it will help them build a conviction. A prosecuting lawyer will ignore facts that make a defendant look innocent. Meanwhile, the defence lawyer will ignore incriminating facts and try to persuade the judge to exclude important witnesses. In the middle of this courtroom tug-of-war hangs the evidence itself, and the forensic scientists who have used all their expertise to produce and interpret it. If it suits the lawyer's narrative, they will undermine first a scientist's testimony and then their good name.

Let's take a typical piece of evidence, the jacket of a murder suspect. As quickly as possible a CSI uses tape to lift any suspect fibres or hairs off the jacket for analysis. Then she puts the jacket in a plastic evidence bag and sends it off to a laboratory scientist who looks for things like bloodstains. After running appropriate tests, the scientist rebags the jacket and stores it ready for a possible appearance as a courtroom exhibit. If the scientist can't find anything useful, the jacket will go into a warehouse to await the next scientific breakthrough that might produce useful evidence, such as extra-sensitive DNA testing.

This is what happened to the jacket belonging to a member of the gang who murdered 18-year-old schoolboy Stephen Lawrence in 1993 in an unprovoked racist attack in south-east London. Stephen was studying for his A-levels, hoping to become an architect. He and a friend were standing at a bus stop in Eltham, on their way home after a night out, when the youths forced him to the ground and stabbed him to death. One member of the gang, Gary Dobson, was wearing a grey bomber jacket. He and his friends always denied murder, although the alibi they gave the police later proved false. Other circumstantial evidence also stood against them, such as footage from a covert camera which the police planted in Dobson's house. Although the gang never discussed the killing on camera, Dobson did call a colleague who took his baseball cap a 'black c***'. When the colleague tapped Dobson on the back of the legs, Dobson said that he unsheathed his Stanley knife and threatened, 'You tap me once more, you silly c***, I'm going to f***ing slice this thing down you.'

Dobson stood trial in 1996 but was acquitted in the absence of physical evidence. However, because of improvements in the sensitivity of forensic tests, and Britain's repeal of the double jeopardy law in 2005 (which means people can now stand trial a second time for the same offence if new evidence has been developed that couldn't have been known at the time of the first trial), the police were ready to launch a major cold case review of the evidence by 2006. They handed the evidence bag containing the bomber jacket to LGC Forensics. And this time scientists found

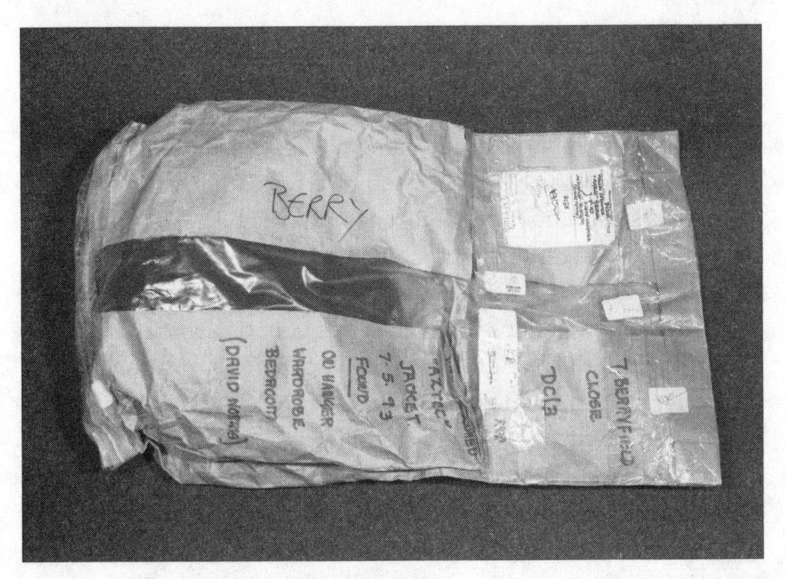

The brown paper bag used to store Gary Dobson's bomber jacket, which was found to be stained with Stephen Lawrence's blood

compelling new microscopic evidence – enough for the police to charge Dobson with the murder again.

At his trial in November 2011, Mark Ellison, the prosecution lawyer, played the video of Dobson's racist rant to the jury, along with footage from the same camera of another member of the gang saying, 'I'd go down Catford and places like that, I'm telling you now, with two sub-machine-guns and I'm telling ya, I'd take one of them, skin the black c*** alive, mate, torture him, set him alight ... I'd blow their two legs and arms off and say, "Go on, you can swim home now."' Ellison also called eyewitnesses to recount the murder for the jury. But the crux of his case against Dobson was what Edward Jarman of LGC Forensics had found on his bomber jacket.

Jarman had spent two days examining the jacket with a microscope and discovered a tiny bloodstain, half a centimetre across, inside the weave of the collar. The jury heard that, after DNA tests, Jarman felt there to be a less than one in a billion chance that the blood belonged to someone other than Stephen.

The stain had been made by fresh, wet blood either from Stephen's knife wounds or from the knife itself. Jarman had also found several dried blood flakes belonging to Stephen at the bottom of the evidence bag. Trapped inside some of these were fibres from the jacket and polo shirt that Stephen had worn on the night of his death. And he found more fibres matching Stephen's clothes when he re-examined the sticky tape that had been used to lift them off Dobson's jacket after the murder.

The journey that physical evidence goes on, from crime scene to courtroom, doesn't get many newspaper column inches. But a trial is the ultimate test of forensic evidence: if isn't well documented, the evidence will not stand up in court. And if it fails in court, all the efforts of the forensic scientists – digital analysts, pathologists, entomologists, fingerprint experts, toxicologists – have been for nothing. At Dobson's trial the defence lawyer, Tim Roberts, made huge efforts to introduce doubt about the journey of the jacket. In his opening statement to the jury Roberts said, 'The charge brought against Gary Dobson is based on unreliable evidence. At the time that Stephen Lawrence was being attacked, Gary Dobson was at home in his parents' house. He is innocent of this charge. The papers in this case are voluminous, but the actual physical evidence, upon which this charge is brought, the fibres and fragments, would not fill a teaspoon.'

For eighteen years the jacket had lain dormant in a paper bag, sealed with tape. Roberts pointed out that in the early 1990s suspect and victim exhibits were often stored in the same room. Over the years many scientists had examined the Stephen Lawrence exhibits in different laboratories around England, and not all of them had worn white plastic suits. Roberts argued that the stain on the jacket collar had not come from fresh blood. He said that the dried blood flakes had got into the evidence bag via a careless scientist who had also handled Stephen's exhibit. He suggested the bloodstain was the result of one of these flakes dissolving when scientists conducted a test for saliva. The test involved wetting and pressing the jacket. Edward Jarman countered that he had tried this theory out using control blood flakes. They had

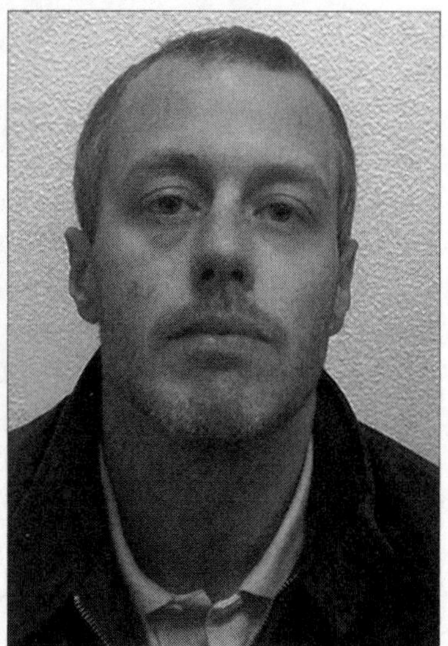

Gary Dobson and David Norris, who were both convicted of the murder of Stephen Lawrence in 2012

turned 'gel-like'; too viscous to absorb into the fabric. The arguments were extensive and detailed. One journalist who was present for every second of Dobson's six-week trial said, 'During these long days of continuity evidence at the end of November – as the barristers argued at length about the safety of brown paper bags – the jury looked bored.'

Roberts was also desperate to get the judge to rule the next witness, Rosalind Hammon, out of testifying. LGC had given her the job of examining the jacket's chain of continuity. He argued that, because she was an employee of LGC, her view could not be trusted. The judge disagreed and allowed her to speak. She testified that, despite the jacket's complex journey, there was 'no realistic probability' the blood and fibres were the result of contamination. On 3 January 2012, Gary Dobson was found guilty of murder and sentenced to at least fifteen years in jail. He had remained out of the reach of justice for eighteen years and 256 days – that's thirty-five more days than Stephen Lawrence was alive. Writer Brian Cathcart said after the trial, 'The idea that there might one day be a conviction was at one point nothing but a fantasy. It's remarkable that you can go back into the evidence and find these microscopic particles which can lead to a conviction.' One other member of the gang, David Norris, was brought to justice at the same trial as Gary Dobson, largely because of a single strand of Stephen's hair found on the jeans that he had been wearing on the night of the murder. Eyewitness accounts suggest that there were three or four other men involved in the murder. Their names are known, but no forensic traces have been found to link them to the scene.

In the Stephen Lawrence case, hard work from the prosecution revealed evidence that led to the conviction of two vicious racist murderers. Evidence that 'would not fill a teaspoon' sank Gary Dobson, with CCTV footage helping him on his way down. But evidence can be a double-edged sword, and sometimes the lawyers can find ways to use it that don't serve justice well.

Jurors love CCTV because, while a lot of evidence used in trials can be argued one way or the other, video evidence provides a clear and irrefutable picture of what happened. But this impartiality means it records *everything* that happened, not just what the prosecution would like to see. In his 2010 book *Defending the Guilty*, criminal lawyer Alex McBride describes a case in which he got someone off through extraordinary use of CCTV evidence. 'Giles' had been caught on a high-definition CCTV camera punching a man in the face. He was stripped down to his waist, and anyone could read his lips in the seconds after the punch: 'Do you fucking want some more?'

As McBride watched the footage to prepare his defence, his heart sank. Despondently he watched to the end of the tape, until it went black. Then suddenly it flickered into life again. McBride watched in astonishment as a police officer pinned Giles's co-defendant, 'Dave', up against a wall, picked him up by his shirt and slammed him to the floor. When Dave's girlfriend tried to intervene, the officer slammed her down, too. She tried to get up, but the officer kept her down with his boot.

McBride showed the clip to Dave's defence lawyer and together they came up with a plan, to demand that all charges against their clients be dropped in return for their not pressing charges against the police for wrongful arrest and assualt. To their delight, the prosecution agreed and Dave and Giles went free. To get his client off, McBride had used an outtake from an otherwise crystal clear moving picture of guilt. 'The golden rule of defence,' he writes, 'is that the less evidence there is, the better – unless it is evidence that contradicts what the prosecution witnesses are claiming under oath.'

Prosecutors serve the state. Their links with the police give them a head start on the defence, but the prosecution must share all its findings with the other side before a trial. The principle of sharing evidence has to do with the legal concept of 'due process', whereby the prosecution may not withhold information that would help the defence. Without due process a fair trial is impossible.

In theory, both the prosecution and the defendant should have access to their own expert witness to give a view on what the evidence means. In particular in US civil courts, the pressure on juries to sort out battling expert opinion has come under criticism. However, UK judges are increasingly encouraging prosecution and defence experts to get together once they have analysed the evidence, to discuss their findings in pre-trial meetings. This saves time, and money which can then be used to pay for other experts. Some US jurisdictions are adopting a similar 'open file policy' in which prosecutors and defence attorneys meet prior to trial to review evidence. As one UK forensic psychologist explains: 'I submit a report, the other side submit a report. If there is a lot of difference we get together with some coffee, hammer out our differences and come to some positive results. It saves both of us going into court for three days and boring the pants off juries who have no idea about the different theories.'

Forensic anthropologist Sue Black agrees. 'It's important to meet beforehand and iron out where the agreements and disagreements are. It helps to remove a lot of the posturing that goes on in the courtroom.' In a recent case in which Sue acted as a defence expert without a pre-trial meeting, it turned into a 'train wreck for the prosecution experts from beginning to end'. At one point the judge asked if the experts from both sides could get together to talk. But the defence and prosecution lawyers both thought there was so little common ground that there would be no point to them meeting. The prosecution's case collapsed and folded, which 'did no one any good at all'.

Experts don't always have to testify in person: whether it's written by one or two experts, a paper report is often enough for the court. Blood expert Val Tomlinson 'deal[s] with far more cases than I'd ever dream of going to court with ... I actually stand in the witness box maybe two or three times a year.' The experience of going on the stand brings up all sorts of emotions – excitement, pride, satisfaction, fear, irritation, humiliation. The mixture depends on the nature of the case – and the personality of the expert themselves.

The very best laboratory scientist may not be able to show the self-control and confidence to perform well in the witness box. Pathologist Dick Shepherd says, 'Lots of scientists can find evidence but there is a unique skill to standing up in court and giving that evidence in a way that a jury with no previous knowledge can understand.' It's a point often made that court proceedings are a kind of theatre – and, as such, good performers, like the famously charismatic pathologist Bernard Spilsbury, often make the best impressions on juries.

While experts are only allowed to answer the questions that lawyers ask them, they are at the same time encouraged to give their opinion. It is their job to find and interpret facts, not to parrot known ones. Of course, the distinction between fact and opinion is a tricky one, and there is a heavy responsibility on the expert witness to say nothing that might mislead the jury. If an expert testifies that a badly smudged fingerprint belongs to Joe Bloggs, is that a fact or an opinion? Or if a blood spatter expert says that the pattern of blood droplets means the victim must have been lying on the ground when the fatal blow was struck, how can a jury assess that evidence?

Furthermore, by its nature, science is provisional: theories are open to rejection or modification in the light of new evidence. Fiona Raitt says, 'Much expert testimony goes to the core of scientific development, which is in a constant state of discovery and refinement. What we know today is sometimes very different from what we knew yesterday.'

Expert testimony is defined as that which is outside the general knowledge of an average member of the public. No matter how strongly they feel about it, the expert witness must leave 'the ultimate question' of guilt or innocence to the jury. To an extent this is a matter of semantics. While Val Tomlinson (see p.155) couldn't say, 'The DNA evidence proves that the Reed brothers did it,' she could (and did) say, 'It is my opinion that the most likely explanation for the DNA results obtained is that the knives were brought to the victim's home by Terence Reed and David Reed respectively, and that they were handling these knives at the time at which the handles broke.'

The principle of 'knowledge beyond the general' was cemented in 1975, after the trial of Terence Turner. Turner was sitting in his car with his girlfriend Wendy, who he thought was carrying his baby. But they argued and, in the heat of anger, she told him that she'd been sleeping with other men while he'd been in prison. One of them had made her pregnant, not him. Boiling over with rage, Turner seized a hammer that was by the driver's seat and smashed Wendy fifteen times across the head and face. He then got out of the car and walked to a nearby farmhouse, where he told someone he had just killed his girlfriend. In court, he said he didn't know what he had been doing, his hand just fell on the hammer and 'it was never in my mind to do her any harm'.

Turner's defence was provocation. If the jury had gone with it, its verdict would have been manslaughter. Instead it found him guilty of murder. He appealed against the verdict on the grounds that the judge had not allowed the jury to hear the report of a psychiatrist. The psychiatrist had written that Turner had no mental illness but was very sensitive to the feelings of others. His 'personality structure' made him vulnerable to anger. And his anger was understandable because of his relationship with the victim. If her confession had taken him by surprise, he could have killed her in 'an explosive release of blind rage'.

His lawyer argued that, if the jurors had been allowed to hear the report, they would have been better able to understand Turner's actions. Lord Justice Lawton reminded the Court of Appeal, 'Jurors do not need psychiatrists to tell them how ordinary folk who are not suffering from any mental illness are likely to react to the stresses and strains of life.' If psychiatrists and psychologists could be called in all cases to prove the probability that the accused was telling the truth, he said, 'Trial by psychiatrist would be likely to take the place of trial by jury.' Turner's appeal was rejected. Fiona Raitt explains: 'The expert has to demonstrate that their field is deserving of the label "expertise" – handwriting is clearly one, so is knowledge of what explosives do – but when it comes to human behaviour, that's the one the judges always splutter at.'

In the vast majority of cases, forensic scientists give the jury important information to consider and help them to understand it. It's the trials that go wrong that give judges and academics like Fiona most to ponder. They are painful for all concerned, but they also pave the way for better justice to be delivered in the next similar trial. Experts with shelf-loads of published work and a Scrabble board of acronyms at the end of their names have the biggest burden of expectation in court. On the whole juries will give their opinion extra weight, especially when they have charisma on their side, too.

A recent example is Roy Meadow, a paediatrician famous for classifying Münchausen Syndrome by Proxy, whereby parents harm their children in order to get the attention of doctors. But in Britain, Meadow's name is best known in relation to Sudden

Infant Death Syndrome, or 'crib death', in which an apparently healthy baby dies with no obvious medical cause. According to Meadow, 'One sudden infant death is a tragedy, two is suspicious, and three is murder until proven otherwise.' Social workers and child protection agencies took 'Meadow's Law' straight to heart, with catastrophic results for a number of families. Similar ideas were in the air in America.

In 1996, an 11-week-old baby boy died suddenly in his bassinet at home in Cheshire. Two years later, his brother Harry died in similar circumstances at the age of only eight weeks. Pathologists found signs of trauma on the babies' bodies. Their mother, Sally Clark, the daughter of a policeman, was arrested and charged with two counts of murder.

Sally stood trial in November 1999. Several paediatricians testified that the babies had probably died from natural causes, believing that the trauma on their bodies had come from attempts to resuscitate them. But the prosecution lawyer portrayed Sally as a 'lonely drunk' who missed her well-paid job as a solicitor and resented her children for keeping her at home. His experts, including Sir Roy Meadow, first thought the babies had been shaken to death, although some later decided they had been smothered. Meadow put the odds against two crib deaths occurring in one affluent household at 73 million to one. He used an analogy to ram the message home: 'It's like backing an 80–1 outsider in the Grand National four years running, and winning each time.' On the back of these very damning stats from a newly knighted doctor, the jury found Sally Clark guilty of murder, by a majority of ten to two.

Sally appealed against her conviction, after the Royal Statistical Society called Meadow's 73 million to one a 'serious statistical error'. To get his figure Meadow had simply squared the 8,500 to one ratio of live births to crib deaths in affluent non-smoking families. This failed to take into account the fact that a sibling of a crib death baby shares very similar genetics and environment, and is therefore at far greater risk than the rest of the population of crib death themselves. The Foundation for the Study of Infant Deaths stated that in the UK second crib deaths in the same family actually occur 'roughly once a year'. But, in October 2000, Sally's appeal was dismissed when the judges claimed that Meadow's figures were a 'sideshow' that would not have affected the jury's decision.

Then, new evidence from Macclesfield Hospital came to light showing that another expert witness at the trial, pathologist Alan Williams, had failed to disclose the results of tests he had done on blood samples. These suggested that one of the babies had died from the bacterial infection Staphylococcus aureus, not from being shaken or smothered. Sally appealed again. This time, in January 2003, her conviction was quashed and she was freed. The appeal court judges commented that, while Meadow had got his stats grossly wrong, it was the findings from Macclesfield Hospital – produced by a lawyer who had worked for free – that had made them overturn the conviction. They lambasted as 'completely out of line' Alan Williams' explanation of why he had discounted the blood test result: because they were inconsistent with his belief that the baby boy had not died from natural causes.

Sally's release prompted a review of shaken baby cases. Two other women, Donna Anthony and Angela Cannings, had their murder convictions overturned and were released from prison. Three of Cannings' babies had died before reaching twenty weeks. She had appealed when it was discovered that her paternal grandmother had had two crib death babies, and her paternal great-grandmother had lost a baby, too. Trupti Patel, who was also accused of murdering her three children, was acquitted in June 2003. In each case, Sir Roy Meadow had testified about the unlikelihood of multiple crib deaths in a single family. 'In general,' he had said, 'sudden and unexpected death does not run in families.'

Roy Meadow and Alan Williams were subsequently struck off the register of the General Medical Council for 'serious professional misconduct'. In 2006, Meadow won his appeal to be reinstated, on the grounds that he had made his statistical mistake in good faith. But, unlike Bernard Spilsbury, his reputation had been tarnished within his own lifetime. In 2009, Meadow himself applied to be removed from the GMC register, meaning he can no

Roy Meadow arrives at the General Medical Council to face a professional conduct committee over evidence he gave in several baby death cases

longer work as a doctor in the UK or testify as an expert witness. British courts no longer prosecute parents of babies suffering crib deaths on the evidence of a single expert witness.

Sally Clark never recovered from her ordeal. Not only had she lost her two tiny sons, but she had been portrayed in the media as a child killer, and then spent three years in prison being treated by other inmates as the epitome of evil. She died in 2007 from alcohol poisoning, at the age of forty-two, leaving her third son without a mother.

Scientists like to give their theories a good airing. Using them successfully in a criminal case improves their standing in the academic community. Sue Black has learned to be wary of this. 'I've been a witness for the court where both sides have agreed that whatever it is that I come out with they will go with it. I'm not saying that that's good, because in certain circumstances, depending upon your expert witness, it could be dangerous.' Pondering over the kinds of subtle errors that are sometimes made, Fiona Raitt

Sally Clark outside the High Court after her release

wonders 'to what extent some experts can be bought. You'd like to think not but it is a very unpleasant world.'

There are clearly dangers in accepting what an expert says at face value. But there is also a danger that the court can go too far the other way and dismiss all cutting-edge science as newfangled and unreliable. In the ideal scenario, judges and lawyers put pioneering scientists under pressure in the witness box, testing the limitations of a technique and giving them new directions to explore back in the lab. When Sue Black first tried to identify a child abuser by the pattern of the veins on his hand, the accused man's lawyer lambasted her for using an unprecedented technique. Anxious that her use of the technique had been a contributing factor in the defendant's acquittal, Sue knew that she still needed to flesh out her vein pattern analysis with more data (see p.185). Ultimately, the technique was used to convict a paedophile who had filmed his abuse of young girls.

This is a good example of how cross-examination can strengthen forensic techniques, by putting pressure on them. If the evidence is sound, the theory goes, jurors will find it all the sounder once they've seen it flexed. But it doesn't always work like that. In fact, people have been questioning the truth-seeking value of the trial system for a very long time. Shortly before his death in 1592, the French lawyer and philosopher Michel de Montaigne wrote, 'First we feel enmity for the arguments and then for the men ... The result of each side's refuting the other is that the fruit of our debates is the destruction and annihilation of truth.' In other words, when lawyers fail in their efforts to attack the evidence, they turn on the person who's giving it instead. One forensic scientist I talked to positively relishes it. 'I love the challenge of being cross-examined by barristers – in the beginning it was, "Young man, in your limited experience ..." Nowadays I often have great fun with barristers because of the gamesmanship.' Robert Forrest is stoical about it: 'If you don't like the heat, you should get out of the kitchen.'

Criminal lawyer Leo Seelig believes it would be 'a great shame' if experts did not have to face the heat of cross-examination: 'To throw into doubt the qualifications of an expert witness is a perfectly reasonable line of enquiry. But it's risky, because it can be unpleasant for the jury and, if you lose them, you've lost the case. The advice given to advocates is not to take on experts at their own game but to put them in the position where, by their own analysis, they can't be sure. And experts do make mistakes all the time. Little things in their reports can be blown up out of proportion. That simply does work, I'm afraid. Which is why experts have to be doubly careful with every detail of their reports.'

In one of Leo's cases, an articulated truck had overturned on a roundabout, fully laden with human sewage which spewed out all over the road. The driver claimed he had tipped because the road had an 'adverse camber', meaning it sloped away from the roundabout, and because another driver had caused him to swerve by driving badly. An expert for the prosecution had used complicated physics to analyse readings from the truck's tachograph – a device which automatically records information relating to speed and distance travelled – to try and prove that the other driver had been driving at a reasonable speed. She backed up her calculations with time-lapsed CCTV footage from a camera that pointed at the roundabout. By the time she had finished answering the prosecution lawyer's questions, the jury was impressed. Then it was Leo's turn to talk to her. '[U]nder cross-examination, her cogent testimony crumbled. She had failed to take into account that the CCTV camera was not at ninety degrees to the lorry, which threw her speed measurements out. By the end she wasn't sure of anything.' Because he had done his homework, Leo was able to draw out an important variable which the prosecution expert had not thought about in the quiet of her office. The other driver was duly acquitted of dangerous driving.

If they're to stand a chance, experts need to show that they deserve the title. One pathologist has noticed lawyers getting keener to catch him out over the years, and thinks that perhaps it has gone too far. 'As I look back over my career there used to be more understanding that an expert witness was there to give you the benefit of the whole of their knowledge, but now we have to reference things. You can no longer say, "Well, look, I've seen twenty cases like this and I think it follows," because the answer is, "Oh, have you published this? Where is your peer-review journal article? You might have got it wrong twenty times mightn't you?" If I say, "I've been doing this job for thirty years and I've seen 25,000 examinations and I've not seen this before," they'll say, "That's just a random event."'

All of the experts I have talked to for this book are seasoned courtroom witnesses. Val Tomlinson has lost count of the number of times she has appeared in court in her thirty-year career – 'hundreds, probably. It can be very intimidating. I remember one case where a lad had died after being been kicked by a number of youths. And another lad had quite a bit of blood in various areas on his trainers but it was difficult to see because it was mixed with cider and was very bubbly. And obviously one of the inferences barristers like to make is that, "There is no blood on my client, and therefore he didn't do it." So I was asked about these particular trainers and I gave my discourse about them being bloodstained. As I walked down the steps from the witness box I was feeling fine. Then the barrister said to me, "Oh, Ms Tomlinson, when you were talking about those trainers I meant to ask you to show them to the jury but I didn't want to interrupt. Would you mind just showing them now?" So the trainers were produced and I walked over to the jury and stood in front of them. I started saying, "You can't see the blood very well, but it's around here," and the barrister behind me erupted. I had finished giving my evidence, and been dismissed from the court. I shouldn't have even been speaking to the jury. There was wafting of gowns going on behind me and I looked at the judge and he said, "Just show them, Ms Tomlinson." So I stood there in front of the jury and I remember raising my eyebrows a little bit, wondering what was going on behind me. When I got outside I sat in the car and I just thought, what on earth happened in there? The display that went on behind me was just ridiculous. Clearly the barrister wanted me to stand there with all the show, and the jury to go, "Well, we can't see anything."'

For forensic entomologist Martin Hall, cross-examination is 'always a nervous time. The heart beats a bit quicker. Your professional opinion is called into question ... You are under intense scrutiny.' What fingerprint expert Catherine Tweedy hates most is 'not being asked the correct questions and not being able to discuss the evidence. You have to sit there and wait for the question and you are not allowed to expand on things. You can at times but a lot of the time you can't. And obviously the opposition try and stop you doing that at every corner because they don't want you to get your point across ... They may entirely miss the point, or simply ignore it even when you know that it's extremely relevant. You don't have control over what is shown to the jury.'

It took our pathologist time to see the adversarial courtroom for what it really was. 'I have only just really realised, when you are giving evidence in court the counsel for the prosecution and the counsel for the defence are involved in legal advocacy, they are not seeking after truth in any way, shape or form. You are sworn to tell the truth, the whole truth and nothing but the truth. Their role is to form an argument and if a bit of what you say is counter to that argument, they will either attack you for it or simply ignore it.'

Witnesses can only tell the 'whole truth' to the questions that they are asked. If they want to tell a truth more whole than that, they run into trouble. As one scientist says, 'It's very difficult as an expert to say, "Excuse me, you have forgotten something." I have done it a couple of times in trials and the look you get from the judge and the barristers – it's not "Oh! Well done, old chap, jolly good, we'd forgotten that, damn it!" Instead, the judge says, "Oh well, I suppose we'll have to look at that then, won't we?" thinking to himself, "Why is this person in my court causing all this trouble? We were doing really well, everyone was singing from his own hymn sheet, until this imbecile piped up." Then you get slapped around by everyone for forty-five minutes, wave the white flag and retreat home.'

Sue Black recognises the courtroom as a potentially 'very rewarding' place, but overall she finds it 'very much the least fun part of the job because it's not our rules. It's not our game. It's the reason why a lot of experts choose to leave the profession because, as you go in, your reputation as an academic is all you have, and it seems at times that part of our adversarial system is a real attempt to rob you of that reputation. It can get very personal. It can get very aggressive. And you will come out of court either still as the world's expert or the world's biggest fool, and I've been both …

'In a recent case my young colleague stood up to give his evidence and was asked, "What is your relationship with Professor Black?" He said, "She's my Head of Department." The defence lawyer replied, "Oh, I think it's a bit more than that, isn't it?" He said to me later he could feel his ears getting red because of the way it was said. It was salacious. He told the lawyer, "I don't know what you mean," and he replied, "Well, I put it to you that she was your PhD supervisor." He said, "Yes," and the lawyer went on, "I put it to you that the professor with her monumental ego looked around her empire and her eyes fell on her favourite little PhD student and she crooked her finger and said, "Do you fancy a day out at the mortuary?" That's how it went, isn't it?" And, bless him, he turned around and said, "No, it bloody well isn't!"

'When it turns to that sort of personal attack, the only disservice that's done is to justice because experts are going to say, "I'm not going to put up with that." And I was very close this year to thinking, why would I do this, why would I keep putting myself through this?'

It's not just seasoned pros like Sue Black, or diligent young experts like her colleague, who are subject to character assassination. A good lawyer will always look for the weakest link in a case – and sometimes that is the victim. A Canadian defence lawyer once gave his colleagues a savage piece of advice. 'If you destroy the complainant in a prosecution ... you destroy the head. You cut off the head of the ... case and the case is dead.'

Fiona Raitt has worked with Rape Crisis, helping victims of rape or sexual assault take their cases to court. For the sake of due process, the defence lawyer of an accused rapist must have access to all of the same medical records of the complainant as the prosecution lawyer, 'Women are shocked when they find out,' explains Fiona. 'They think, "How did they get that?" The defence lawyer will say, "Is it the case that you were on tablets, let's see, oh, tranquillisers, about three years ago because you had a bout of mental health issues?" And before you know it they are creating a story of this un-credible person who probably can't remember very well and who perhaps is still taking tablets. For one reason or another, the most vulnerable witnesses are the ones with the longest medical records, and the defence have a field day. Complainants do have the right to refuse to pass over their records but they often don't because they haven't properly grasped the significance of revealing them.' What medical information the defence has access to in the US is restricted by the

Health Insurance Portability and Accountability Act, and also varies widely from judge to judge and hospital to hospital.

In January 2013, Frances Andrade, a violinist, was the complainant in the trial of Michael Brewer, her old music teacher, whom she had accused of rape and indecent assault. On the stand, she was repeatedly called a liar, and she was reduced to tears by the cross-questioning. In a text message to a friend, she wrote that the experience of giving evidence was 'like being raped all over again'. Less than a week after giving testimony, and before the end of the trial, she killed herself at her home in Guildford, Surrey. Brewer was convicted of five counts of indecent assault.

When Louise Ellison, Professor of Law at the University of Leeds, set up a mock jury of forty members of the local community, and got actors and barristers to re-enact rape trials in front of them, she found that jurors were influenced by the demeanour of the complainant in court – whether emotional or composed – and by how long after the rape the complainant had reported it. When a judge or an expert explained how varied reactions to an unwanted sexual approach could be, however, the jury was less likely to be put off giving a guilty verdict either because of a calm demeanour or a delay in reporting the rape.

But the default position for a judge is to keep quiet, explains Fiona. 'There are cases where judges haven't intervened even though the witness has been weeping, collapsed in the witness box in tears. They've said, "We'll take a short break, and can somebody get her some water?" They try to do nothing which indicates that they are not impartial. They have to be very careful. But … I think they could actually protect witnesses a great deal more than they do.' Judges have to be careful about intervening because if it seems even slightly that they are taking sides, the verdict of the trial can be overturned on appeal.

The idea that jurors should be left to make up their own minds is the cornerstone of any adversarial criminal justice system. But there is almost no research testing this principle in practice with real live juries. Academics like Fiona Raitt and Louise Ellison are not allowed to do any research on real juries to see what they make of the evidence and argument shown to them. Ellison's study raises the question, would a judge with experience of working with rape victims be better placed to pass judgment than jurors plucked from the general public?

Other factors make the courtroom a difficult place for the jury, too. No studies have been done on their ability to balance complex forensic evidence presented to them over a trial that may last several weeks. Fiona even remembers a time when 'jurors weren't allowed to take notepads in because they were supposed to watch what was going on at all times'. Some jurors must be left in a state of confusion by the new concepts scientists teach them, by barristers' attempts to dismantle those concepts, and by statements from other scientists which contradict them. Juries do not always get it right, and they do give the wrong weight to certain evidence. A 2014 study by legal experts and statisticians from Michigan and Pennsylvania found that 4.1 per cent of prisoners sentenced to death in America were innocent.

Some people find the process of cross-examination so unhelpful that they would like to do away with it altogether. As opposed to the adversarial system used in Britain and America, many countries, like France and Italy, use a combination of jury trials and the inquisitorial system in which, rather than lawyers presenting opposing sides of the argument, a judge investigates the facts of a case. The judge questions witnesses and the accused (or their lawyer) before the trial and only if she finds enough evidence of guilt does she call for a trial. At that point she hands over all the evidence she has gathered to the prosecution and defence lawyers. At the trial she may question witnesses again, to clarify what they said in their pre-trial testimony. The prosecution and defence lawyers are not allowed to cross-examine witnesses, but they are allowed to present the jury with a summary of their views.

There are benefits and disadvantages to both systems. Trial by jury has its roots in ancient Greece and Rome, and began in England in 1219. As their powers increased, the jury came to be seen as a pillar of society: a group of your equals could condemn you to jail but a wigged member of the establishment could not. By the eighteenth century the law recognised that juries were there to limit the state's ability to lock up the people it didn't like. Italy's trial of American Amanda Knox has highlighted both the difficulties of forensics in a contaminated crime scene, and the limitations of the inquisitorial system.

Doing away with juries has been tried before in the Diplock Courts in Northern Ireland, set up in 1973 during the Troubles to stop the harassment of jurors. Some people think that the Diplock judges sitting in isolation got it right more than they got it wrong, and more often than juries got it right. And the Diplock model is, in Fiona's words, 'quicker, much quicker' – which is important when you think about the thousands of pounds it costs to run a court every day. But Michel de Montaigne, again, has some pertinent thoughts on this system of justice: 'A judge may leave home suffering from the gout, jealous, or incensed by a thieving valet: his entire soul is coloured and drunk with anger: we cannot doubt that his judgment is biased towards wrath.'

Leo Seelig defends the adversarial system. 'The real beauty of the adversarial system is that, so long as both parties are competent, then you really do hammer out all the issues, and they are properly litigated. There is an ethos amongst defence counsels in the UK that you must fight your case fearlessly and justly.' From the point of view of the scientists, an inquisitorial system would end the 'posturing' and aggressive character assassination that they so loathe. Nevertheless, some of them would be against such a radical change. It's useful to remember what Peter Arnold said at the beginning of this book. 'I actually see the need for an adversarial system. I was challenged but ultimately that strengthened the case because it was clear that there were no issues. Ten years down the line we're not going to have an appeal in that case saying the evidence could have been tampered with. I'd rather get it out in the open now. Let's challenge it now. Let's face the scrutiny.'

Other scientists think that the kind of intense scrutiny lawyers currently put on them would be better directed. As one says,

'I've had a defence solicitor in my office say to me, "Well, you know, we know he's guilty as sin but it's our job to catch you out." That's the thing that offends me more than anything. No, it's not their job to catch us out. Their job is to look at the evidence.'

In the experience of a fire expert I talked to, 'The court process is a game between the lawyers and the experts. Lawyers may misinterpret the very best science that you put in front of them, and deliver a different message to the jury.' Similarly, Fiona Raitt sees the same mismatch between the pursuits of the adversarial system and the pursuit of truth: 'I don't think those who defend the adversarial process believe that it is the best way to get at the truth … I think it distorts the truth actually. There is deep reluctance for governments to explore what juries do. It is probably too terrifying, because they'll discover that actually they are highly prejudicial. A lot of those prejudices come out of the way that they deliberate. Basically it's the strongest juror that wins the day and everyone else just falls into place.'

The British exported adversarial trials and the jury system around the Empire. They remain the way of justice in countries such as the US, Canada, Australia and New Zealand. The US is the country best known for its adversarial system, partly because cameras are often allowed into the courtroom. Even more than in the UK, competent lawyers and less ethical expert witnesses in American courts can be swayed by the high fees they can command. The best illustration is the all-star legal team assembled by O. J. Simpson in 1995, to defend him against the charge of stabbing his wife, Nicole Brown Simpson, and another man, Ronald Goldman, to death.

In that infamous trial, lead defence lawyer Johnnie Cochran helped to get the jury on side with a mix of Technicolor suits, sharp cross-examination and burning charisma. At one point the prosecution asked Simpson to put on a glove that had been recovered from his house and which was – according to their case – bathed in the victims' blood and Simpson's own DNA. In court Simpson found it difficult to put the glove on. Cochran cocked his head to the jury and said, 'If it don't fit, you must acquit!' The prosecution suggested that the glove had shrunk because it had been frozen and unfrozen several times during DNA testing. They produced a photo of Simpson wearing the glove some months before the murder. But neither the glove nor a slew of other incriminating evidence was enough to stop O. J. Simpson walking free, though he was later found criminally liable by a jury in a civil trial brought by the Brown and Goldman families.

More commonly, the accused is not a wealthy sports star. When it comes to hiring lawyers and experts, most people have to settle for what they can afford. The civil rights campaigner Clive Stafford-Smith's book *Injustice* (2013) follows the extraordinary case of Krishna 'Kris' Maharaj, a British businessman convicted of a double murder in a hotel room in Miami in 1986. The jury found Kris guilty of killing his Jamaican business partner, Derrick Moo Young, and his son, Duane Moo Young. Now seventy-five, Kris has so far spent twenty-seven years in a Florida jail for the crime.

At the trial, prosecution lawyer John Kastrenakes made a powerful opening statement to the jury: 'You will hear scientific evidence regarding fingerprints, ballistics evidence, business records ... All of it points to this defendant – nobody else – as the killer.' Kris's fingerprints were found in the hotel room where the murders had been committed – because, Kris said, he had attended a business meeting there earlier in the day. Kastrenakes called an array of eyewitnesses and experts, including a police officer who testified to selling Kris a 9 mm Smith & Wesson handgun some months before the murder. Kastrenakes made a compelling case, studded with phrases like 'mechanically planned', 'brutal act' and 'overwhelming evidence'.

When it was the turn of the defence lawyer, Eric Hendon, to call on his witnesses, he shocked all present. He simply said, 'The defence rests'. Hendon had six people ready to confirm that Kris had been with them at the time of the murder, in a location forty miles away from the hotel. But the jury never heard from them. Incomprehensibly, Hendon completely blew his chance to introduce doubt into Kastrenakes' narrative.

The jury deliberated for a short time and found Kris guilty of first-degree murder, causing him to faint in his chair. Later on, the same jury returned to the courtroom and sentenced him to death.

For innocent murder suspects to have a lawyer of Hendon's calibre is not uncommon in the US. By definition the guiltless are unlikely to know much about what the criminal justice system requires of them: they feel their innocence will speak for itself. Keen to clear their name, they rush to trial without getting a competent team together to trouble the prosecution's case. Kris had paid Hendon a flat fee of $20,000. (By comparison, O. J. Simpson had spent around $10 million on his defence team, which works out at $16,000 per expert per day.) In Stafford-Smith's words, 'Capital punishment means those without the capital get the punishment.' As for experts, Kris couldn't bear to spend the money. What was the need to refute evidence that couldn't possibly exist? Even though he had made money importing fruit from the Caribbean into the UK, Kris would eventually bankrupt himself and his long-suffering wife Marita in the appeals process.

It may have been that there was more to Hendon's lacklustre performance than lack of monetary incentive. According to the investigator who worked with him on Kris's defence, he had received a threatening phone call a few weeks before the trial. Something would happen to his son, the caller said, if it looked like he was doing too much to get Kris off.

The prosecution's case had relied on more than just the healthy fee and bustling energy of lead man John Kastrenakes. The witnesses played their part, particularly ballistics expert Thomas Quirk. The question of the murder weapon was at the front of the jury's mind, since the police had never found it. Quirk testified that the bullets found in the bodies of the Moo Youngs had been fired from one of six possible brands of 9 mm semi-automatic. He had fired all these possibilities in his lab, and found that the marks on the bullets – made by the spiralled 'rifling' on the inside of the gun barrels – resembled the marks on the fatal bullets.

Quirk then talked about the bullet casings which the CSIs had recovered from the hotel room: 'The only fired standard that

I have in the lab that matches the morphology on the casings from the scene is a Model 39 Smith & Wesson.' Considering a police officer had already told the court that Kris had bought this very gun some months prior to the murder, these were damning words.

Finally Quirk presented a photograph of a silver Smith & Wesson handgun to the jury. It filled the gap left by the absent murder weapon, and stuck in their minds. Hendon objected to Quirk showing the weapon, saying it had nothing to do with the facts, but the judge snapped,'It's demonstrative!' and let him go on. When Hendon cross-examined Quirk, he was able to get him to admit that around 270,000 Smith & Wesson handguns had been produced in the US since the 1950s, and that the bullet might have been fired from any one of them. But by this point the jury had in a sense seen the murder weapon.

Was Quirk's science valid? Could he really narrow down the source of the bullets to a Model 39 Smith & Wesson? Or had the Moo Youngs been killed by another of the 65 million handguns knocking around the US in 1986? The ability of ballistics experts to match a bullet to a gun – 'ballistics fingerprinting' – had not been properly challenged since its inception in the nineteenth century. Like fingerprint examiners and forensic hair specialists, ballistics experts had been reluctant to question the scientific basis of their own livelihood. It was only in 2008 that Jed Rakoff, a federal judge in New York, finally held hearings to look into the status of ballistics evidence. He suggested that it had been more reliable in the days when bullets were made from individual moulds, but was much less so in the era of mass production. 'Whatever else ballistics can be called,' he said, 'it cannot fairly be called "science".' In the wake of Rakoff's study, some American jurisdictions will no longer allow ballistics experts to testify to a conclusive match, and defence attorneys have much to work with in challenging ballistics evidence.

After the trial it came to light that Quirk regularly gave testimony in the language of absolute certainty. In the murder trial of Dieter Riechmann, for example, accused of killing his girlfriend in the front seat of his rental car in Miami Beach in October 1987, he testified that the fatal bullet had come from one of three types of gun, two of which Riechmann owned. Riechmann was subsequently convicted and sentenced to death. At an appeal hearing ten years later, Quirk admitted that he had run the details of the bullet only through the Miami Police database, rather than through the FBI database which had many thousands more possibilities.

Via his charity, Reprieve, Stafford-Smith has been investigating the Moo Young murders for a decade. He has brought a great deal of new evidence to the surface, from the police files and from the people involved in the case.

In the hotel room where they were shot, the Moo Youngs had documents detailing their laundering of as much as $5 billion for the notoriously violent Medellín drug cartel in Colombia. They were trying to skim off 1 per cent for themselves, which might have annoyed the cartel. Most significantly, the original jury was never told about the person in the hotel room opposite the Moo Youngs' – a Colombian who was under investigation for hiding $40 million in his luggage en route to Switzerland. No other guests were staying on that floor of the hotel on the day of the murders.

In 2002, Kris's sentence was reduced to life, with the possibility of parole when he reaches the age of 103. In April 2014, a Miami judge granted Kris a full evidentiary hearing on the strength of the new evidence. According to Reprieve, 'this represents the biggest step towards Kris's exoneration since his conviction in 1987'.

In the adversarial system, a fair trial is impossible without access to the same resources, what under UK law is known as 'equality of arms.' At the very least Kris Maharaj should have had a good lawyer and a ballistics expert. If any theory is to be trusted – whether it is a theory of guilt or anything else – it needs to be scrutinised and criticised by competent outsiders. The scientific method demands it.

Without the scrutiny of the courtroom, the science assembled by forensic experts is meaningless. The job of forensic science is to support the legal system from the crime scene to the courtroom. But everything depends on that final stage being scrupulous and even-handed. That is not only in the best interests of science; it's in the best interests of all of us.

CONCLUSION

This book has charted the astonishing leaps that forensic science has made over the last two hundred years. If we presented Michael Faraday or Paracelsus with the scientific evidence our courts now take for granted, it would seem like magic to those most rigorous of researchers. And the advance of science has run hand in hand with corresponding advances in the delivery of justice.

When beat officer John Neil arrived at the scene of the first Jack the Ripper murder in 1888, he faced insurmountable problems. No one in Whitechapel's intricate network of alleyways and streets had seen the murderer on that August night. There was no obvious motive, and no obvious suspect. Mary Nichols's body offered evidence about the murder weapon, about the strength of the murderer himself and the state of his twisted mind. But none of this pointed in any decisive direction.

Had Neil and his colleagues had the skills and technology of modern forensic investigators, processing the scene would almost certainly have led them to follow Holmes' 'scarlet thread of murder' inexorably to the man who killed those Whitechapel women in the dead of night. But without the most basic of scientific resources, the police were fumbling in the dark. They knew it, and the public knew it: a popular cartoon of the time showed a blindfolded officer stumbling hopelessly around in a street full of Rippers laughing and goading him.

The five acknowledged victims of the Ripper were Mary Ann Nichols, Annie Chapman, Elizabeth Stride, Catherine Eddowes and Mary Jane Kelly. They represent a tiny proportion of the men, women and children whose killers have escaped retribution, simply because there was no way to unravel the complex circumstances of a murder scene. But the police and forensic services have learned lessons from these failures that ultimately have served to protect others. Even the several thousand dogs who died slow deaths by poison at the hands of the 'father of toxicology', Mathieu Orfila, in the early 1800s, had a significant role to play.

In the course of researching this book, I have been struck, above all, by the integrity, ingenuity and generosity of the forensic scientists I have met. They care so deeply about the cases they work on that they are willing to engage with the darkest and most frightening aspects of human behaviour on a daily basis. They are willing, like Niamh Nic Daéid , to spend hours in the sodden debris of a fatal fire; like Martin Hall, to collect maggots from a week-old corpse; or, like Caroline Wilkinson, to reconstruct the face of a mutilated child the same age as her own daughter. They make sacrifices so that the rest of us can live knowing that, if we are the victims of crime, the perpetrators will be brought to justice. They do not guard their knowledge jealously; they share it as widely as possible in the hope that one of their colleagues may use it as a springboard for the next leap forward.

And the importance of their work makes them astonishingly creative in the face of a tricky forensic problem. The proliferation of forensic tools made available to crime investigators over the last two hundred years is nothing short of astounding. And, although they are all imperfect, nearly all have strengthened the criminal justice system. We've heard about the 'bucket science' that characterised the early days of DNA analysis; now a scientist like Val Tomlinson or Gill Tully can use a bloodstain a millionth the size of a grain of salt to provide a profile that can find not only the person it belongs to, but also a member of their family who might have committed a crime, maybe years ago. Confronted with a video that appeared to show sexual abuse but not the abuser's face, Sue Black became the first person to identify someone from the unique pattern of veins on their forearms and freckles on their hand. These scientists find their imaginations stimulated rather than curbed by the challenges of crime investigation – and by the need to be rigorous.

Crime scene evidence would not be so effectively utilised today if it hadn't, for more than two hundred years, been forced to pass the strict credibility tests of the courtroom. The first pressure is put on a scientist's theory by their scientific peers, who force them either to abandon it or meet the challenge and make it stronger. Then, in the courtroom, lawyers do everything they can to excite scepticism in the jury. Very few holds are barred on the witness stand, and a lawyer may choose to ignore their scientific methods and interrogate their character instead. But, however personally stressful a forensic scientist may find giving testimony, the courtroom is the anvil on which scientific evidence is struck. With a well-prepared lawyer playing the part of the hammer, forensic techniques are either strengthened or broken, according to their merit.

Of course, as parts of this book have shown, it doesn't always run like clockwork. But when it does, inspirational sparks fly, new ideas are knocked out and the room for manoeuvre enjoyed by violent criminals shrinks a little more.

The methods of science and justice have much in common. Both attempt to shine a clarifying light on obscurity and uncertainty. At best, their core aims match, too, as they try to go beyond assumption and arrive at the truth though demonstrable facts. Yet because forensic science is made up of so many human layers – criminals, eyewitnesses, police officers, CSIs, scientists, lawyers, judges, juries – it cannot avoid either missing or misrepresenting the truth at times. The stakes are always high; life and liberty depend on it. I hope this book has demonstrated the commitment of forensic scientists across the disciplines to be imaginative, open-minded and painstakingly honest in the interests of justice for all of us. It has certainly reminded me of what I have known for a long time – the work itself is amazing and the people who do it are, frankly, awesome.

ACKNOWLEDGEMENTS

I was lucky enough to be educated in Scotland, where the education system allows students to study arts and sciences alongside each other all the way up to university level. I enjoyed both equally and I still love being gobsmacked by the latest developments in science and technology.

However, I am primarily a writer of fiction, albeit one with an appetite for authenticity. But when I get stuck, I generally make something up. So when it comes to writing non-fiction, I need a lot of help. Thankfully, it was forthcoming.

In the first instance, I owe a huge debt to the experts I interviewed in the various disciplines I've written about here. It was a privilege to be in touch with their enthusiasm, good humour and insight into what is often challenging and harrowing work. Some I've known and exploited for years; others are new to the experience. I could not have begun to produce this book without their generosity with their time and their expertise. So thank you Peter Arnold, Mike Berry, Sue Black, Niamh Nic Daeid, Robert Forrest, Martin Hall, Angus Marshall, Fiona Raitt, Dick Shepherd, Val Tomlinson, Gill Tully, Catherine Tweedy and Caroline Wilkinson. And for helping with the vagaries of American law, Patrick Hoffman and Brigid Hoffman.

I've had enormous support and help right from the start of this project from Kirty Topiwala and her colleagues at the

Wellcome Trust, backing me up with a wide range of facilities – from Bernard Spilsbury's handwritten notes to all the coffee I could drink!

I had two first-class researchers who supplied me with exactly what I needed along the way. Anne Baker and Ned Pennant Rea were both patient and efficient. I couldn't have written the book without their help. Nevertheless, I take full responsibility for any errors.

Most of all, I want to thank publisher Andrew Franklin at Profile, who first came up with this crazy idea, and my editor Cecily Gayford who has gone the extra mile so many times, she's covered the equivalent of the London Marathon. I can't believe you never shouted at me. I would have.

Finally, thanks to my indefatigable agent, Jane Gregory, who always has my back, and to my family, who are always there for me when I need them.

SELECT BIBLIOGRAPHY

Arthur Appleton, *Mary Ann Cotton: Her Story and Trial* (London: Michael Joseph, 1973)

Bill Bass, *Death's Acre: Inside the Legendary 'Body Farm'* (London: Time Warner, 2004)

Colin Beavan, *Fingerprints: The Origins of Crime Detection and the Murder Case that Launched Forensic Science* (New York: Hyperion, 2002)

Carl Berg, *The Sadist: An Account of the Crimes of Peter Kürten* (London: William Heinemann, 1945)

Sue Black & Eilidh Ferguson, eds., *Forensic Anthropology: 2000 to 2010* (London: Taylor & Francis, 2011)

Paul Britton, *The Jigsaw Man: The Remarkable Career of Britain's Foremost Criminal Psychologist* (London: Bantam Press, 1997)

David Canter, *Criminal Shadows: Inside the Mind of the Serial Killer* (London: HarperCollins, 1994)

David Canter, *Forensic Psychology: A Very Short Introduction* (Oxford: Oxford University Press, 2010)

David Canter, *Forensic Psychology for Dummies* (Chichester: John Wiley, 2012)

David Canter, *Mapping Murder: The Secrets of Geographical Profiling* (London: Virgin Books, 2007)

David Canter & Donna Youngs, *Investigative Psychology: Offender Profiling and the Analysis of Criminal Action* (Chichester: John Wiley, 2009)

Paul Chambers, *Body 115: The Mystery of the Last Victim of the King's Cross Fire* (Chichester: John Wiley, 2007)

Dominick Dunne, *Justice: Crimes, Trials and Punishments* (London: Time Warner, 2001)

Zakaria Erzinçlioğlu, *Forensics: Crime Scene Investigations from Murder to Global Terrorism* (London: Carlton Books, 2006)

Zakaria Erzinçlioğlu, *Maggots, Murder and Men: Memories and Reflections of a Forensic Entomologist* (Colchester: Harley Books, 2000)

Colin Evans, *The Father of Forensics: How Sir Bernard Spilsbury Invented Modern CSI* (Thriplow: Icon Books, 2008)

Stewart Evans & Donald Rumbelow, *Jack the Ripper: Scotland Yard Investigates* (Stroud: History Press, 2010)

Nicholas Faith, *Blaze: The Forensics of Fire* (London: Channel 4, 1999)

James Fallon, *The Psychopath Inside: A Neuroscientist's Personal Journey into the Dark Side of the Brain* (London: Current, 2013)

Roxana Ferllini, *Silent Witness: How Forensic Anthropology is Used to Solve the World's Toughest Crimes* (Willowdale, Ont.: Firefly Books, 2002)

Neil Fetherstonhaugh & Tony McCullagh, *They Never Came Home: The Stardust Story* (Dublin: Merlin, 2001)

Patricia Frank & Alice Ottoboni, *The Dose Makes the Poison: A Plain-language Guide to Toxicology* (Oxford: Wiley-Blackwell, 2011)

Jim Fraser, *Forensic Science: A Very Short Introduction* (Oxford: Oxford University Press, 2010)

Jim Fraser & Robin Williams, eds., *The Handbook of Forensic Science* (Cullompton: Willan, 2009)

Ngaire Genge, *The Forensic Casebook: The Science of Crime Scene Investigation* (London: Ebury Press, 2004)

Hans Gross, *Criminal Investigation: A Practical Handbook for Magistrates, Police Officers, and Lawyers* (London: Sweet & Maxwell, 5th edition, 1962)

Neil Hanson, *The Dreadful Judgement: The True Story of the Great Fire of London, 1666* (London: Doubleday & Co., 2001)

Lorraine Hopping, *Crime Scene Science: Autopsies & Bone Detectives* (Tunbridge Wells: Ticktock, 2007)

David Icove & John DeHaan, *Forensic Fire Scene Reconstruction* (London: Prentice Hall, 2nd edition, 2009)

Frank James, *Michael Faraday: A Very Short Introduction* (Oxford: Oxford University Press, 2010)

Gerald Lambourne, *The Fingerprint Story* (London: Harrap, 1984)

John Lentini, *Scientific Protocols for Fire Investigation* (Boca Raton: CRC Press, 2013)

Douglas P. Lyle, *Forensics for Dummies* (Chichester: John Wiley, 2004)

Michael Lynch, *Truth Machine: The Contentious History of DNA Fingerprinting* (Chicago, London: University of Chicago Press, 2008)

Mary Manhein, *The Bone Lady: Life as a Forensic Anthropologist* (Baton Rouge: Louisiana State University Press, 1999)

Mary Manhein, *Bone Remains: Cold Cases in Forensic Anthropology* (Baton Rouge: Louisiana State University Press, 2013)

Mary Manhein, *Trial of Bones: More Cases from the Files of a Forensic Anthropologist* (Baton Rouge: Louisiana State University Press, 2005)

Alex McBride, *Defending the Guilty: Truth and Lies in the Criminal Courtroom* (London: Viking, 2010)

William Murray, *Serial Killers* (Eastbourne: Canary Press, 2009)

Niamh Nic Daéid, ed., *Fifty Years of Forensic Science: a commentary* (Oxford: Wiley-Blackwell, 2010)

Niamh Nic Daéid, ed., *Fire Investigation* (New York: Taylor & Francis, 2004)

Roy Porter, *The Greatest Benefit to Mankind: A Medical History of Humanity from Antiquity to the Present* (London: HarperCollins, 1997)

John Prag & Richard Neave, *Making Faces: Using Forensic and Archaeological Evidence* (London: British Museum Press, 1997)

Fiona Raitt, *Evidence: Principles, Policy and Practice* (Edinburgh: Thomson W. Green, 2008)

Kalipatnapu Rao, *Forensic Toxicology: Medico-legal Case Studies* (Boca Raton: CRC Press, 2012)

Mike Redmayne, *Expert Evidence and Criminal Justice* (Oxford: Oxford University Press, 2001)

Mary Roach, *Stiff: The Curious Lives of Human Cadavers* (London: Viking, 2003)

Jane Robins, *The Magnificent Spilsbury and the Case of the Brides in the Bath* (London: John Murray, 2010)

Andrew Rose, *Lethal Witness: Sir Bernard Spilsbury, Honorary Pathologist* (Stroud: Sutton, 2007)

Edith Saunders, *The Mystery of Marie Lafarge* (London: Clerke & Cockeran, 1951)

Keith Simpson, *Forty Years of Murder* (London: Panther, 1980)

Kenneth Smith, *A Manual of Forensic Entomology* (London: Trustees of the British Museum (Natural History), 1986)

Clive Stafford-Smith, *Injustice: Life and Death in the Courtrooms of America* (London: Harvill Secker, 2012)

Maria Teresa Tersigni-Tarrant and Natalie Shirley, eds, *Forensic Anthropology: An Introduction* (Boca Raton: CRC Press, 2013)

Brent E. Turvey, *Criminal Profiling: An Introduction to Behavioral Science* (Amsterdam; Oxford: Academic Press, 2012)

Francis Wellman, *The Art of Cross-examination: With the Cross-examinations of Important Witnesses in Some Celebrated Cases* (New York: Touchstone Press, 1997)

P. C. White, ed., *Crime Scene to Court: The Essentials of Forensic Science* (Cambridge: Royal Society of Chemistry, 2004)

James Whorton, *The Arsenic Century: How Victorian Britain was Poisoned at Home, Work and Play* (Oxford: Oxford University Press, 2010)

Caroline Wilkinson, *Forensic Facial Reconstruction* (Cambridge: Cambridge University Press, 2008)

Caroline Wilkinson & Christopher Rynn, *Craniofacial Identification* (Cambridge: Cambridge University Press, 2012)

George Wilton, *Fingerprints: Scotland Yard and Henry Faulds* (Edinburgh: W. Green & Son, 1951)

ILLUSTRATION CREDITS

While every effort has been made to contact copyright-holders of illustrations, the author and publishers would be grateful for information about any illustrations where they have been unable to trace them, and would be glad to make amendments in further editions.

1 SCENE OF THE CRIME

3 Police Officer Sharon Beshinevsky. Photo: Getty Images

4 Doctor Edmond Locard, Founder of Police Sientific Laboratory of Lyon. Photo: Maurice Jarnoux/*Paris Match* via Getty Images

16 Detectives combing the area around Sharon Beshinevsky's murder scene for evidence. Photo: Getty Images

2 FIRE SCENE INVESTIGATION

21 Michael Faraday, whose 1861 book *The Chemical History of a Candle* paved the way for modern fire scene investigators. Photo: Wellcome Library, London

25 Fire scene investigators at the scene of the Stardust Disco Fire, in which 48 people died and more than 240 were injured. Photo © *The Irish Times*

30 The fossilised remains of a diatom, a single-celled organism – viewed under a microscope. Photo: Spike Walker/Wellcome Images

3 ENTOMOLOGY

4 PATHOLOGY

5 TOXICOLOGY

6 FINGERPRINTING

119 The Bertillionage record of twenty-one-year-old George Girolami, arrested for fraud. Photo: adoc-photos/Corbis

123 A CID assistant checks a new set of prints against Scotland Yard's fingerprinting records in 1946. Photo: Getty Images

126 Buck Ruxton's fingerprints, taken in Liverpool Prison, 1936. University of Glasgow Archive Services, Department of Forensic Medicine & Science Collection, GB0248 GUAFM2A/25

135 Spanish forensic experts search wreckage after the Madrid train bombings. Photo: Pierre-Philippe Marcou/AFP/Getty Images

7 BLOOD SPATTER AND DNA

141 From top left: Samuel Sheppard after the alleged attack, his wife Marilyn Reese Sheppard and Sheppard testifying at his trial in a neck brace. Photo: Bettmann/Corbis

143 Dr Paul Kirk examines blood spatter on Marilyn Sheppard's pillow. Photo: Bettmann/Corbis

151 Colin Pitchfork, the first person in the UK to be convicted on the basis of DNA evidence. Photo: Rex Features

8 ANTHROPOLOGY

167 Forensic anthropologists excavating a mass grave in Kosovo. Photo: AP/PA Photos

170 Clyde Snow testifies at the 1986 trial of nine former Argentinean military junta. Photo: Daniel Muzio/AFP/Getty Images

172 Members of the Argentine Forensic Anthropology Team excavate a common grave. Photo: EAAF/AFP/Getty Images

9 FACIAL RECONSTRUCTION

195 A collection of 'criminal faces' compiled by Cesare Lombroso. Photo: Mary Evans Picture Library

204 A photograph of Alexander Fallon, a victim of the King's Cross fire, compared with the facial reconstruction created using his remains. Photo: PA Photos

206 The 'Butcher of Bosnia', Radovan Karadžić. Photo: AFP/Getty Images

10 DIGITAL FORENSICS

11 FORENSIC PSYCHOLOGY

12 THE COURTROOM

COLOUR PLATES

1 Crime scene notes taken by John Glaister Junior. University of Glasgow Archive Services, Department of Forensic Medicine & Science Collection, GB0248 GUAFM2A/1

2, 3 Police officers comb the area where the remains of Isabella Ruxton and her maid, Mary Rogerson, were found. University of Glasgow Archive Services, Department of Forensic Medicine & Science Collection, GB0248 GUAFM2A/73 and 109

4 A maggot's head under a microscope. Photo: Science Photo Library/ Getty

5 A blowfly feeding on decaying meat. Photo: Wikimedia Commons

6 An illustration from Eduard Piotrowski's seminal work on bloodstains

7 A body in situ at the 'Body Farm', Tennessee. © Sally Mann. Courtesy of the Gagosian Gallery

8, 9, 10 Graham Coutts, who was convicted of Jane Longhurst's murder, caught on CCTV moving her body from the storage facility where he kept it in the weeks after her death. Photos: Rex Features

11, 12, 13 Death of a Court Lady, from a series of Japanese watercolour paintings, *c.* 18th century. Wellcome Library, London

14, 15 Betty P. Gatliff works on a facial reconstruction. Photo: PA Photos

16 Sections of a brain, showing bullet path and bullet. Image courtesy of Bart's Pathology Museum, Queen Mary University of London

17 Section of liver, showing knife wound and knife. Image courtesy of Bart's Pathology Museum, Queen Mary University of London

18 One of Frances Glessner Lee's 'Nutshell Studies of Unexplained Death'. Courtesy of Bethlehem Heritage Society/The Rocks Estate/ SPNHF, Bethlehem, New Hampshire

19 A model of old man's head in wax, created by the seventeeth-century sculptor Giulio Zumbo. Bridgeman Art

INDEX

Page references for photographs are in *italics*